# ROUTLEDGE LIBRARY EDTIONS: GLOBAL TRANSPORT PLANNING

Volume 18

# THE RURAL TRANSPORT PROBLEM

# THE RURAL TRANSPORT PROBLEM

DAVID ST. JOHN THOMAS

LONDON AND NEW YORK

First published in 1963 by Routledge & Kegan Paul Ltd

This edition first published in 2021
by Routledge
2 Park Square, Milton Park, Abingdon, Oxon OX14 4RN

and by Routledge
605 Third Avenue, New York, NY 10017

*Routledge is an imprint of the Taylor & Francis Group, an informa business*

© 1963 David St. John Thomas

All rights reserved. No part of this book may be reprinted or reproduced or utilised in any form or by any electronic, mechanical, or other means, now known or hereafter invented, including photocopying and recording, or in any information storage or retrieval system, without permission in writing from the publishers.

*Trademark notice*: Product or corporate names may be trademarks or registered trademarks, and are used only for identification and explanation without intent to infringe.

*British Library Cataloguing in Publication Data*
A catalogue record for this book is available from the British Library

ISBN: 978-0-367-69870-6 (Set)
ISBN: 978-1-00-316032-8 (Set) (ebk)
ISBN: 978-0-367-74887-6 (Volume 18) (hbk)
ISBN: 978-0-367-74895-1 (Volume 18) (pbk)
ISBN: 978-1-00-316009-0 (Volume 18) (ebk)

**Publisher's Note**
The publisher has gone to great lengths to ensure the quality of this reprint but points out that some imperfections in the original copies may be apparent.

**Disclaimer**
The publisher has made every effort to trace copyright holders and would welcome correspondence from those they have been unable to trace.

# THE RURAL TRANSPORT
PROBLEM

by
DAVID ST. JOHN THOMAS

LONDON
ROUTLEDGE & KEGAN PAUL

*First published 1963
by Routledge & Kegan Paul Ltd
Broadway House, 68–74 Carter Lane
London, E.C.4*

*Printed in Great Britain
by Latimer, Trend & Co Ltd, Plymouth*

© *David St. John Thomas 1963*

*No part of this book may be reproduced
in any form without permission from
the publisher, except for the quotation
of brief passages in criticism*

*By the Same Author*

A REGIONAL HISTORY OF THE RAILWAYS OF
GREAT BRITAIN, VOL. I: THE WEST COUNTRY
RURAL TRANSPORT: A REPORT
LAKE DISTRICT TRANSPORT REPORT
GREAT MOMENTS WITH TRAINS
TRAINS WORK LIKE THIS
THE MOTOR REVOLUTION

CONTENTS

Introduction *page* xi

PART ONE

1. Inconvenience or Hardship? 3

2. The Historical Background 11
   The Railways—Bus Services

3. Railway Management 21
   Waste in the West—The Burden of the Past—Dieselization — Train Services — Signalling — General Administration—Conclusion

4. When a Branch Line is Closed 36
   The Unnecessary Loss of Public Goodwill—The Transport Users' Consultative Committees — Where the Committees Have Failed

5. Bus Services 45
   An Over-confident Industry—The Provision of Rural Services—Self-knowledge and Public Relations — Road and Rail Co-operation — Small Operators, Tendering and Subsidy—The Traffic Commissioners—Miscellaneous Points

6. Transport and Rural Life 65
   The Cost of Avoiding Subsidy — Conclusions Reached from Research Work — Planning and Transport

7. Subsidy and Other Administration 75
   Recommendations in Reports, 1960–62 — A Professional and a Voluntary Committee — The Cost of Subsidy

v

*Contents*

## PART TWO

8. Research Work                                               *page* 89
    The Village Surveys—The Closure of Two Branch Railways—A Study of Bus Passengers—Tourist Traffic

9. Case Histories                                          126
    1. The Teign Valley—2. Newton Abbot–Moretonhampstead—3. Coniston—4. The Keswick Line—5. Mid-Northumberland Bus Operators

Postscript                                                       166
    The Beeching Plan—Other Railway Developments—The North Devon Railway Inquiry—Bus Matters

Index                                                                   173

# TABLES

### VILLAGE SURVEYS

1. General Summary                          *page* 96
2. Ownership of motor vehicles                96
3. Occupation of vehicle owners               97
4. Daily journeys                                  101
5. Frequency of travel, by destination most visited    102
6. Places regularly visited                       103
7. Use of public transport                        104
8. Transport difficulties                           107
9. Women who do not personally own cars; numbers who drive in car-owning households    109
10. Reasons for not driving given by non-car owners living in car-owning households    109

### DEVON BRANCH LINE SURVEYS

11. Effects on former users of closure of Teign Valley railway    112
12. Effects on former users of closure of Moretonhampstead branch    115

### LAKE DISTRICT BUS SURVEY

13. Residents' combined summer and winter figures; frequency of travel by bus    118
14. Residents' summer and winter figures; purposes of travel    119
15. Visitors' summer figures; frequency of local travel during holidays    120
16. Visitors' summer figures; means of arrival in Lake District    121

*Tables*

### LAKE DISTRICT VISITORS' SURVEY

17. Cross-section sample of those staying in the Lake District *page* 124
18. Probable effects upon visitors of closing branch lines in the towns concerned  124
19. Means used by visitors staying in the Lake District for travel during holiday  125

### MID-NORTHUMBERLAND BUS OPERATORS

20. Passengers, miles and receipts of four operators  159

# MAPS

The Lake District showing the survey villages of the
Lake District Transport Inquiry  *page* 90

The Teign Valley and Moretonhampstead areas  130

Mid-Northumberland Rural Transport Survey Area.
Frequency of Bus Services, 1951  161

Mid-Northumberland Rural Transport Survey Area.
Frequency of Bus Services, 1960  162

Mid-Northumberland Rural Transport Survey Area.
Frequency of Bus Services, forecast for 1965  163

# INTRODUCTION

THIS is the first full-length book on the problem of communications in rural Britain. It has two chief aims: to demonstrate the importance, human and economic, of preserving some passenger public transport in country areas even at the cost of subsidy; and to discuss whether the resources available for train and bus services could not be used more effectively than at present.

The book is in two parts. The first deals generally with the problem and possible solutions—the habits and needs of country people and of those who provide their transport. The second is devoted entirely to an outline of the original research made under the auspices of the Dartington Hall Trustees and of the Lake District Transport Inquiry. Many of the facts quoted in the first part were of course yielded by this original research.

Some readers will of course disagree with my conclusions. I am, indeed, reluctant to put forward fresh proposals, so numerous and varied have been other people's ideas in recent years. Even the members of the Committee on Rural Bus Services set up by the Government failed to agree among themselves, and their majority report proved equally unacceptable to the Government and to the transport industry. Committees sitting in Scotland and in Wales reached different conclusions on many matters.

My suggestions have been made carefully in the light of the reports of these committees and the reactions to them, and also of course of my own research programme, the largest yet completed on the subject. It is, however, chiefly the hard facts produced by the research work, rather than the recommendations, which will, I hope, make this book a useful contribution to rural literature and sociology. I hope that these facts might be studied

*Introduction*

even by those hotly opposed to the kind of solutions which seem to me best.

Hitherto there has been a serious lack of knowledge. As was stated by John Saville in his introduction to *Rural Depopulation* in this series, there is in Britain no general body of rural sociologists, and a large proportion of the writings on rural problems are the work of agricultural economists. Had one-tenth of the cost of the agricultural research and advisory services been bestowed on country transport, there is not the least doubt that the situation would be far brighter today.

One of my chief contentions is that transport operators themselves do not always fully understand their business, and often fail to serve their own best interests as well as those of the public. Deciding the extent to which it is justifiable to criticize transport management has been difficult. Sometimes it is held that estimates of the savings from branch-line closures, for example, are the domestic concern of British Railways, and should not be queried. Having seen the problems in close-up, I have concluded that an assessment of operating methods is relevant.

Something should perhaps be said of the history of this book. The project was commissioned by the Dartington Hall Trustees in December 1958. As so little basic data was available on the subject, it was essential to carry out widespread investigations. Simultaneously with the collection of information in the country as a whole, I undertook two more detailed surveys, in Mid-Northumberland and the Teign Valley of Devon.

While the work was taking place, the Government set up the Committee on Rural Bus Services with Professor D. T. Jack as chairman. As it was not possible to publish this book in time for the Committee's consideration, my interim findings were set forth in March 1960 in a booklet called *Rural Transport: A Report*.[1] The same month another move occurred which has delayed the completion of this volume, though strengthening its factual foundations. Sir Patrick Hamilton, Chairman of the North-Western Transport Users' Consultative Committee, had been growing increasingly concerned about the lack of adequate background information against which to consider applications

[1] David & Charles, Dawlish, 1960.

*Introduction*

made by British Railways to close uneconomic branch lines. With the support of his committee, he therefore established an Inquiry to study the public transport problem in one rural area of the North-West—the Lake District.

The Lake District Transport Inquiry was a unique venture. It was entirely independent, but had some financial backing from the three county councils and the Lakes Planning Board, and the goodwill of the Transport Commission, the railways and bus companies, and of many local organizations. I was appointed director, and with a staff of eight working in the area for two study periods it was possible to collect a large amount of detailed information. The results of the Inquiry were published in May 1961 in *Lake District Transport Report*.[1] The most important information is repeated here, and comparisons are made for the first time with the Northumberland and Devon investigations.

A large number of people have contributed to making this book possible. My especial thanks are due for the help and encouragement of the Trustees of Dartington Hall, and of their secretary, Mr Peter Sutcliffe, and to Sir Patrick Hamilton, who generously allowed the material collected in the Lake District to be used here. Mr. Cyril Dove, secretary of the North-Western T.U.C.C. and of the Lake District Transport Inquiry, also provided much invaluable help. Mr. J. R. Shepherd, of the National Institute of Social and Economic Research, has taken part in the research schemes in Devon, Northumberland and the Lake District, and is responsible for editing the important statistical side of the book. Mr. Douglas Mennear took part in the Northumberland surveys and has kept me in touch with later developments in the area. Among those who have also provided material or answered inquiries are Mr. D. S. Barrie, Mr. C. R. Clinker, Mr. B. Grocott, Mr. Ivor Hocking, Mr. G. O. Holt, Mr. B. S. Murton, Mr. T. E. Opie, Mr. N. T. O'Reilly, Rev. A. Harold Page, Mr. Alec Trotter and Mr. P. Turnbull. It is, of course, impossible to list the transport officials who so kindly gave their time to answering my queries, and thanks are due too to all those hundreds of country dwellers who patiently answered questionnaires.

[1] David & Charles, Dawlish, 1961.

*Introduction*

Mr. Victor Bonham-Carter has given me great encouragement in the actual writing of this book, while Mr. J. A. B. Hibbs has kindly read and commented on the principal sections on bus services, and Mr. G. Mills on Chapter Four.

Finally, my wife has taken an active part in all stages of the work and added much to the enjoyment of the task.

DAVID ST. JOHN THOMAS

NOTE. The text has not been revised since completion in 1963, but later developments are considered in the Postscript.

# PART ONE

CHAPTER ONE

# INCONVENIENCE OR HARDSHIP?

THIS book is written with considerable feeling. In the course of my research not only have I continuously met people whose enjoyment of life has been sapped by inadequate transport, but time and again it has seemed that much of the inconvenience to the rural population (and also much of the damage to the rural economy) is unnecessary.

The average standard of transport management in the countryside is, with outstanding exceptions, disconcertingly low. In many cases a more useful service could be provided at far less cost. Too frequently buses and trains are run with scant reference to the demand. Narrowly-missed connections are perpetuated in the timetables year after year, sometimes decade after decade. The railways in particular are often guilty of useless expenditure; fully-staffed signal-boxes may survive long after the removal of any possible justification for them.

Inevitably a study of the rural transport problem leads first to criticism of the operators, who could have done so much more to help themselves and the public. Sometimes indeed these men seem almost to enjoy the martyrdom of watching losses mount to the point where closure is inescapable. But examination of the difficulties in their wider context brings sympathy for railway and bus officials on the job and the realization that the policy-makers are largely to blame.

The chief trouble has perhaps been the lack of foresight on the part of transport—especially railway—management. It was not realized in time that the railways could not pay their way and continue to provide extensive social services. If the Transport Commission had presented the Government with a forceful, clear-cut alternative between policies, probably some decision would have been reached early enough to prevent much of the damage. On the other hand, the Government need not have evaded taking the initiative in giving the industry the lead it so obviously needed.

Throughout the 1950's, arguments continued inside and outside the industry as to whether the railways should be regarded as a public utility or whether they should be run on strictly business lines disregarding the interests of minorities. Although it became increasingly clear toward the end of the decade that basically the Government insisted that the system should pay its way, loopholes remained, the question of the retention of social services was shunned even in the 1960 White Paper, and many railwaymen felt that 'justice' for British Railways still lay round the corner.

The result was a prolonged and extremely unhappy compromise. Although many services had been closed by the time British Railways entered a new era under the management of Dr. Richard Beeching in 1962, thousands of miles of uneconomic routes were retained—including some branch lines which British Railways wished to close but could not because they were unable to meet the condition then in force that adequate bus alternatives had to be provided. Yet because so many routes had no assured future—closure was always potentially imminent—improvements and even short-term working economies were shelved. In addition, capital was not available for schemes on many lines more likely to have a future. With few exceptions, it was impossible for railwaymen at district headquarters to obtain sanction to spend even small sums of capital on schemes which would have yielded valuable annual savings. Some lines which could have been brought near to paying their way ten years ago had thus become hopelessly uneconomic by 1962.

Much the same thing has happened with rural bus services. Many country routes have always run at a loss and have been 'cross-subsidized' by profitable town and inter-town business. But by 1952 it was clear that without Government help there could be no long-term security for services in more sparsely-populated regions. By 1955 warnings were being uttered by some Members of Parliament and others that unless subsidies were introduced many parts of Britain would be left without public passenger transport. Not until December 1958, however, was the Jack Committee appointed to report on the position. The Committee's findings were published in July 1960, a delay in itself harassing to operators uncertain whether or not to prune services further, and two years later the Government had still not indicated any intentions.

Exchequer help, should it eventually arrive, may be too late to save many important services. Without help, the bus industry will be in far worse plight than had it known from the beginning that it would have to fend for itself, whatever the consequences to its customers. Obviously while Government aid has been a possibility, the public have been in a strong position to demand postponement of withdrawals. Unfortunately, the relatively small number of withdrawals in the four years following the Jack Committee's appointment has sometimes been quoted as evidence that there is no financial problem, the commercially-minded urban world wrongly assuming that if buses remain on the road they must be profitable. Alas, I have spent long hours examining the balance sheets of small operators whose capital has steadily disappeared and who remain in business solely to avoid letting down the public while a shred of hope for better times persists. After years of indecision, it seems that anything more than token support for rural transport is unlikely.

Rightly or wrongly, intentionally or unintentionally, steps have been taken to keep an active and prosperous rural economy in Britain. Agriculture is directly subsidized, and also indirectly supported by the expensive agricultural advisory service and its ancillaries. Roads, power, water, education, postal and telephone services all cost more than country people are expected to pay for them.

The country dweller—farmer, farmworker, and retired townsman who wants to spend his last days in peaceful surroundings—benefits in three ways. Firstly, many charging systems are inevitably standardized: posting a letter costs 3d. anywhere, for example. Secondly, some organizations have thought it prudent to serve the countryside at less than cost even where this could be avoided: if they wished, the electricity boards *could* reduce the extent to which sparsely-populated areas are a drag on town consumers. And thirdly, many social and other services (including education, library and health) are made available to the countryman as a right. They are of course provided as cheaply as possible, but they are provided whatever the cost.

Why omit transport, thereby vitiating the value of some of those services? Three arguments are used in answer to this question.

1. It is popularly held that transport operators, like the Post Office, should be prepared to take the rough with the smooth, making sufficient profit on main routes to cover at least the essential rural ones. This opinion persists despite clear contradiction by those who have inquired into the matter—the Select Committee on Nationalized Industries,[1] for example. Today transport is among the most competitive of industries. If they raise charges on main routes, operators immediately lose business, either to other men who concentrate on profitable runs and incur no rural losses, or to the ubiquitous private car—including the hired car. Only if operators with rural liabilities were fully protected on their main routes—and today that is impossible—would the argument about 'using the fat to fry the lean' hold true. Of course there is still some fat: the territorial bus companies continue to support thousands of miles of unprofitable country routes out of urban and inter-urban takings, but this becomes steadily more difficult. Moreover, fat and lean territories are distributed indiscriminately.

To be quite clear, this is essentially a matter of degree. The Post Office would be able to reduce telephone tariffs by connecting only subscribers likely to prove profitable, and no doubt the tariffs high enough to cover rural losses reduce the attraction of the service to town subscribers. But the telephone service is a monopoly. Again, electricity charges in towns reflect the high cost of serving rural areas, and this weakens electricity's competition with other fuels in the town market. But the weakening is slight, and in no way endangers electricity's position. When we come to transport, we find that fares high enough to cover losses on unremunerative services are driving away large slices

[1] Report of the Select Committee on Nationalized Industries: British Railways (Stationery Office, 1960).

of the most valuable traffic on profitable routes. The traveller from London to Plymouth goes by road because he does not wish to pay the main-line railway fare that includes a contribution to the upkeep of secondary lines.

2. It is said that public transport is fast becoming unnecessary in view of the increase in car ownership. This too is untrue. Certainly in some country areas one in three adults now owns a motor vehicle, and public services are less used. But a railway or bus route can still play a vital part in rural life even when only a small proportion of the population use it regularly, and even when receipts cover only a small part of costs. Conclusions drawn from the ratio of costs to receipts can be misleading when assessing the usefulness of a service. That some services affect the lives of far more people than actually use them is a point dealt with fully later in the book.

3. Those who oppose subsidies argue that although cuts in transport bring some inconvenience, this amounts to nothing serious. 'It's just the changing times. A few people are bound to suffer, but it isn't as though there were any real hardship. We'd soon hear about that.' The fact that a growing number of backbenchers from the West Country, the North of England and elsewhere have lost no opportunity to point out the dangers of the present position is overlooked, because the majority of M.P.s are convinced that were the crisis pressing they would hear much more from such organizations as the National Farmers' Union and the county councils. Did the problem directly affect a large number of farmers and farmworkers, there is no doubt that action would have been taken long before this. But the effect is of course indirect and often not fully appreciated. Farmers report that their workers complain of nagging wives who find life 'too quiet', but the analysis goes no further. The vast majority of county councillors are motorists. Even meetings of Women's Institutes attract a higher proportion of car owners and those living on the best transport routes than of other women. The very lack of mobility tends to prevent those who suffer from inadequate transport voicing their grievances. Including many young and old people, they are anyway not a naturally vocal part of the population.

## *Inconvenience or Hardship?*

It is true that taking the narrow view the amount of positive hardship does appear relatively small. Except in isolated cases, the residents of even remote areas obtain the necessaries of life, and the benefits of the Welfare State. Few families have to move because the chief wage earner is unable to reach work. There is nothing dramatic to justify shock headlines, but a great deal of inconvenience is endured quietly by an increasing number of people in many parts of the country. Sometimes this inconvenience is proving severe enough to become an underlying, perhaps only half-realized, source of restlessness in family life. It is this sapping of the contentment of our rural areas that is the real problem. Inevitably it is difficult to describe and does not interest the popular Press.

I have spoken of inconvenience, but where does inconvenience end and hardship begin? It is impossible to draw the boundary but clearly some of the deprivations which one generation have accepted are not going to be accepted by generations to come. To some extent the mere passage of time is turning inconvenient conditions into unbearable conditions. The luxuries of yesterday are the necessities of today. Already a much higher standard of transport is needed to preserve equanimity in more socially advanced parts of the country, such as South-East England, than in less developed areas, such as the central Devon plateauland and the upland country to the east of the Lake District. Having had better education, many daughters cannot face the lack of opportunities their mothers accepted. Indeed, in many cases the mothers, though not complaining for themselves, are determined that their daughters' lot shall be different. 'It's all right for us, but it won't do for the young ones,' were words I frequently heard during interviewing.

The problem of rural transport does of course very largely concern girls and young women. Boys take jobs locally on farms, in forestry or in quarries, or perhaps drive lorries which they park at home at night; if they go to the towns to work, they buy motor vehicles at an early age. But on leaving school many girls are confronted with impossibly difficult journeys to and from the nearest town. Sometimes they board out, if only on alternate nights; in other cases whole families are moving specifically to give the daughters better chances of advanced education, employment and, of course, marriage.[1]

As has already been said, if poor transport were preventing workers reaching the farmers, rapid action could be expected. But most agricultural workers have to take only a short walk or bicycle ride. They need little transport: television, pub and football pools satisfy most of their entertainment demands. It is very different for the wives, especially the younger ones. Inability to enjoy leisurely shopping, to meet friends, and to share even to a limited extent in urban amusements frequently leads to discontent. The result may be years of frustration and bickering in married life. Or the husband may be persuaded to save domestic happiness by moving to town. In this case he will probably be lost to agriculture (urban living demands urban wages and the theory that workers will travel back to the countryside daily does not work), but most farmers will be more inclined to blame the young woman's unreason than even to think of the transport position.

True, some farmers see that times have changed, offer the use of their own car, arrange regular lifts, or encourage staff with young families to live in the village or on the bus route rather than in isolated cottages nearer their work. But this is exceptional. 'I've lost my man because his wife wouldn't bide,' is more familiar, and there is genuine lack of comprehension as to why young women won't settle where their predecessors grew old. 'I've even put in the electric,' a farmer may add to show his puzzlement that his cottage is no longer considered desirable.

The lack of understanding by individual farmers is perhaps not surprising. What is both surprising and tragic is that no responsible authority is aware of the changing outlook—of the standards that will be demanded tomorrow, and the difficulties that will be encountered in meeting them—let alone prepared with plans for action. Especially serious is the neglect of the transport field by town and country planning authorities. Few planners have troubled to assess the importance of a bus as a

[1] I am primarily concerned with the local, working indigenous population: daughters of families in the upper income group usually leave home to pursue a career, but farm workers' wives have neither the means, the traditions nor the inclination to send away their girls at 15.

village amenity: thus in Devon, for example, we find two neighbouring villages, one with a modern sewerage system so that new housing is now encouraged, but without a bus service owing to narrow lanes, and the other with a bus service but a ban on expansion for lack of adequate sewerage.

On the one hand country people will continue to demand a higher standard of living. On the other, it is going to be increasingly difficult financially to maintain public transport in rural districts. What the outcome will be, no one can accurately discern. But it is certain that a more understanding appreciation of the human and economic problems and a determination to make the best of circumstances could soften the blows to the rural economy in the years ahead. It is to encourage such a constructive spirit that this book with its many criticisms of existing arrangements is written.

CHAPTER TWO

# THE HISTORICAL BACKGROUND

### THE RAILWAYS

BRITAIN's economy, rural as well as urban, largely developed round the railway network. With their vast powers of 'creative destruction', the railways, the first form of mechanical transport, established trends which represented as big a break from tradition as anything seen in this country. The full significance of the term Railway Age has perhaps not yet been realized.

The first main-line railway was opened in 1830. A mere thirty years later the system of trunk routes was largely completed and its power demonstrated. New towns were rising, and towns bypassed by the trains were declining. Comings and goings between London or provincial capitals and rural areas had intensified at least a dozenfold. Greenwich time had been universally adopted. National newspapers were being increasingly read and greater space was being allotted to national events by local newspapers, aided by the telegraph, a railway adjunct. Many small mills whose future had seemed assured at the beginning of the generation were being forced to close. In short, the processes of centralization and standardization had been set in motion.

As the century wore on, even in the remotest countryside, at least in England, people were to some extent affected by the trains. Villages lost trade to towns or to larger villages on the railway map. Carriers working to and from market towns also made station connections: if few passengers were transferred, the occasional intercourse with the outside world was of profound importance, and the exchange of goods increased until railway rates were of concern to farmers everywhere, and manufacturers of popular branded goods could achieve a truly national distribution down to the remotest village shop.

## The Historical Background

Not only did villages off the railway map turn ever more often to their railhead station, but by 1914 an enormous number of quite small villages and hamlets—in a few cases even individual farms—had their own wayside station or halt. At its peak Britain's railway system was almost absurdly dense, and though few costings were made it is clear that even before the advent of motors much of it never paid. Many rural lines were built by locally-sponsored companies, which eventually sold out to their big neighbours at a price well below the original cost. Others were built by the large concerns themselves, frequently less to serve the local population than to assist manœuvres against rival railways for the profitable long-distance traffic. Thus some of the branch lines in Mid-Northumberland were originally envisaged as through routes to Scotland. The duplication of routes between towns was often economically unjustified, but each new line did carry trains into a further slice of countryside.

For two generations the railways held an almost complete monopoly. They wielded unparalleled power which, as I have explained in detail for one region elsewhere, they frequently abused.[1] Although there were highlights, especially on the technical side, generally services were improved only when imperative, and there was little voluntary exertion to satisfy the public. Mechanical transport developed as one of the least efficient and most quarrelsome of industries. But from the beginning many railwaymen acknowledged that monopoly involved responsibilities as well as privileges, and though opportunities to help both themselves and their customers were ignored, the tradition of providing some services which would inevitably lose money was established. The railways themselves, of course, were sole judge of what charity should be meted out to a minority of travellers at the cost of the customers at large. From this tradition have stemmed many of the industry's difficulties in the last forty years.

If before 1910 the railways could prosper while being inefficient, tightfisted with most of their customers, and extravagant in rivalry with their neighbours, they could not do so once

[1] *A Regional History of the Railways of Great Britain.* Vol. I: *The West Country* (Phoenix House, 1960).

## The Railways

motor vehicles had proved themselves. But management was slow to realize the seriousness of the challenge—quite unprepared for the invasion of the roads by motor buses after 1918. On some routes the railways lost the cream of local traffic by 1925, but little was done either to fight back or to reduce the size of the system. Pruning was hardly started before 1930, and even then caution prevailed. Only 118 passenger stations were closed on the former London & North Western Railway system and its subsidiaries between 1930 and 1947, compared with 278 between nationalization on 1 January 1948 and the end of 1960.[1]

Undoubtedly, had the management of the four big railway companies of the 1923-47 Grouping Era been able to cast aside past traditions, practices and prejudices, rural dead wood would have been cut far more firmly, to conserve resources for the competition with road transport on the main fronts. But officials still thought, as some do even now, in terms of monopoly, involving certain duties as well as privileges. To run a railway as any other business, ignoring the broader public benefit, was a notion few entertained. Other factors of course limited the effectiveness of railway competition between the wars, but management voluntarily continued to accept the obligation of supplying unremunerative services in thinly-populated areas.

The war and nationalization brought little change in outlook. When the Transport Commission were formed, they were charged with providing 'adequate railway services' throughout the country, the needs of agriculture and the rural population being borne in mind. In 1953, however, the requirement of adequacy was dropped. The Commission were then technically free to run what railway services they chose, subject to the Transport Users' Consultative Committee procedure (described in Chapter 4), and to the two general statutory duties of securing enough revenue to 'make provision for the meeting of charges properly chargeable to revenue, taking one year with another', and having 'due regard ... to the needs of the public, agriculture, commerce and industry'. As the Select Committee on Nationalized Industries pointed out in 1960, these two duties

[1] C. R. Clinker: *London & North Western Railway Chronology, 1900-1960* (David & Charles, Dawlish, 1961).

conflicted. 'In having regard to the public need, the Commission are providing services which detract from their chances of making their revenue match their expenditure.'

Until the end of 1961 the Commission themselves continued to assess the 'public need'. Little fresh blood had been drafted to the top levels of railway management; most district superintendents, for example, had joined the service when it was still conceived in terms of monopoly. Moreover, a high proportion of the executive staff were still in the territory where they had begun their careers. It is therefore not surprising that social obligations should have been regarded seriously, many services covering less than a third of their costs being allowed to survive.

There was another factor. Throughout the 1950's a strong belief persisted among railwaymen at all levels that in the end the Government would recognize the desirability of maintaining the system substantially as it was and grant some kind of general subsidy. That the question of 'social transport' obligations was shunned so long—avoided even in the 1960 White Paper—was a hard blow to officials trying to keep their industry intact. Uncertainty discouraged economies, for most economy schemes involve capital expenditure only justifiable where services are to continue at least for a few years. Yet certainly many worthwhile savings, and adaptations of services to meet changed conditions, could have been made even on a short-term basis. When the staff of the Lake District Transport Inquiry studied the railways of that area in 1960, the impression was gained of an organization lacking in enthusiasm to make the best of the situation. A large proportion of the staff with whom we talked were chiefly concerned to point out missed opportunities.

With the appointment of Dr. Richard Beeching as chairman of the Transport Commission in 1961, a new era began, the railways setting out to run on strictly commercial lines for the first time in their history. A vast programme of research was inaugurated, the railways themselves undertaking fact-finding of a kind which in this country had previously been attempted only by the author. The Minister of Transport, not the Transport Commission, became the judge of social requirements.

During the 1950's many people had thought in terms of a general subsidy to British Railways to cover all social, strategic and other uneconomic requirements. But the writing of a blank cheque had everything against it. A subsidy would almost inevitably be too small to plug mounting losses, or so large that prodigious sums of the taxpayers' money would vanish without adequate control over its results.

It was the author who first suggested that while British Railways as a whole should be expected to run strictly as a business concern, subsidy should be distributed 'in a number of grants for limited, specified purposes, enabling the railways to retain, modernize, and in a few cases perhaps even reopen, branch lines and wayside stations which have a value to the public'.[1] At first this idea of specific subsidies for specific lines met little approval, partly because at that time many people still felt that the railways themselves should remain the assessor of public need, and partly because of the difficulty of producing accurate accounts for one line. Railway accountancy has advanced since then, and the latter difficulty is now largely removed.

### BUS SERVICES

The development of bus services was as dramatic as that of the railways had been over two generations before. The first motor-bus ran in 1898, and the present system was virtually complete by 1930. Although many of the large 'territorial' bus companies originated before 1914, the great expansion was concentrated into the twelve years after 1918.

Operators after the war belonged to four classes. The territorial companies, including many alive today, extended their networks and opened many rural as well as inter-urban routes. Some newcomers to transport arrived, including ex-servicemen who had driven during the war and bought vehicles with their gratuities, and a few people with business knowledge and greater capital resources who saw unlimited opportunities in the field. Village carriers, many with generations of horse-and-cart experience behind them, rapidly adopted the new vehicles. And a few railway companies, notably the Great Western, developed bus routes as adjuncts to their train services.

Expansion was encouraged by three factors. There was no

[1] *Rural Transport: A Report.*

## The Historical Background

statutory control over entry into the industry, technical progress and road improvements made it more attractive, and the demand for transport was brisk. At first the bus was regarded by the railways as ally rather than enemy. As early as 1903 railway companies began experimenting with buses for feeder services, and by 1912 a number of branch line and light railway schemes had been abandoned because it had already become obvious that motors would provide the cheapest short-distance transport in areas of sparse population. The opening up of the countryside by buses stimulated travel and contributed traffic to the longer-distance train services. But as the bus industry expanded, it of course became a serious rival to the railways for medium and long-distance traffic as well as for local trips. The first express coach ran in 1925 and quickly ate into railway traffic, though no doubt also stimulating new business.

By 1927 the G.W.R.'s buses alone were carrying 8,000,000 passengers annually. But this was a stake insufficient for the railways, which clearly stood in danger of losing control of a large proportion of Britain's passenger-carrying business. With the deliberate aim of protecting their capital, in 1928 the four main-line railway companies obtained powers more fully to engage in bus operation themselves. Then, by an important agreement in 1929, the large road combine controlled by the joint undertaking of Tilling and British Electric Traction joined forces with the railways rather than risk a clash of interests. The railways became almost half-owners of most of the chief provincial bus companies, whose strength was greatly increased by the flow of railway capital. The combine as well as the individual companies also grew: the railways contributed their own services, the companies they had already bought, and capital to buy others. With a few minor exceptions, the railways ceased actually to run the buses themselves: railwaymen had not proved born bus operators.

The agreement of 1929 and the subsequent expansion were alike made possible by the Road Traffic Act of 1930, which both parties foresaw, the bill having been in draft in 1928.

Throughout the 1920's, new routes could be opened without licence, and on the more popular runs the multiplicity of operators was as picturesque as it was wasteful. The London

## Bus Services

Traffic Act of 1924 had restricted entry into the capital's passenger-carrying business. The 1930 Road Traffic Act applied to the whole country. It set up the Traffic Commissioners to administer a system of route licensing. This had two important results. Because they were protected, operators with established routes now had a saleable goodwill, which encouraged small concerns to sell and the territorial companies backed by railway capital to buy. And in return for a measure of protection against newcomers on their most profitable routes, operators came to accept the obligation to provide a fair quota of useful services which did not pay—on the principle of 'using the fat to fry the lean'. This principle was not new, but the Act made possible the carrying of services at an outright loss out of monopoly profits guaranteed by statute.

The work of the Traffic Commissioners has changed little during the last generation. Reporting on the licensing of road passenger services in 1953, the Thesiger Committee stated:

> 'Throughout the country a high proportion of unremunerative services has been made possible and operators generally have readily co-operated in meeting the need for such services wherever possible. The Licensing Authority has no direct power to force an operator to put on or keep on a service but in practice operators have recognized an obligation to provide as full as possible a network of services for the area they cover.'[1]

Since 1930, services have been concentrated in the hand of the combines—Tilling, British Electric Traction (the original combine was split) and the Scottish Omnibus Group—and today these have an outright monopoly in parts of Britain.[2]

---

[1] Report of the Committee on the Licensing of Road Passenger Services (1953).

[2] 'It is estimated that in Great Britain as a whole between 70 and 80 per cent of all rural services are provided by companies in the three large groups. There are also a number of large- and medium-sized independent undertakings operating stage services. Many of these services are in rural areas.'—Report of the Committee on Rural Bus Services (1961).

Although this may give a fair over-all impression, it should perhaps be added that a higher proportion of the mileage of the independent operators than of the combines is off the beaten track and exclusively rural in character. Many so-called rural routes of course link towns, and inter-town traffic may account for a large part of the receipts.

## The Historical Background

The detailed story of the combines need not concern us. Briefly, in 1948 the Transport Commission purchased the former Thomas Tilling group. The whole of their capital became railway-owned. For the most part the companies have been free to compete against the railways, although there have been some inhibitions. The Transport Commission also held a substantial, though minority, holding in the British Electric Traction group, whose operation is entirely divorced from railway management, except that again railwaymen are among the members of the Boards. The position is about to change as this book goes to press. Under the 1962 Transport Act, the railways' business interests are transferred not to the new Railway Board but to a Transport Holding Company, entirely separated from the railways. It is not yet possible to forecast in detail how the change will work, although there are serious apprehensions among both bus and railway interests.

The bus industry continued to expand both during and immediately after the war. With wives at work and many men away from their homes, the need for travel increased. The food shortage encouraged frequent visits to shops. Cinemas still provided the chief entertainment for many families, and holidays had to be taken in Britain. With higher wages, people were able to afford more journeys, and turned to the buses because cars were scarce and expensive and petrol was rationed. Independent operators played a considerable part in wartime and post-war development, including the expansion of rural services and the establishment of town local services and better connections between rural areas and seaside resorts. Fares were held down until about 1951.

The causes of decline in bus patronage are well known. The number of private cars licensed in Great Britain doubled between 1950 and 1958, the increase being even more rapid than this in some rural areas. The habit of giving lifts to neighbours has become stronger with the corresponding decline in country bus services. Although many townspeople owning cars have moved into the country, the indigenous rural population has fallen in most areas. Television has largely replaced trips to the cinema. Increased delivery services, mobile shops and libraries allow fewer journeys to the nearest town: often the market-day

## Bus Services

visit is continued, and not others. The easing of hire-purchase restrictions, and the generally more plentiful supply of consumer goods between 1951 and 1955 meant that many people had to cut current expenditure: this was one reason for the interruption in the growth of holiday resort business, and indeed partly accounts for the first railway deficit in 1956. The almost yearly increases in bus fares since 1951 have of course met consumer resistance. Another factor, sometimes overlooked, is that stage-service buses running to a published timetable have faced increasing competition from excursion and private-hire services, which are not subject to the same licensing restrictions.

When the decline actually began in rural areas is not easily proved. In the nation as a whole the number of passengers reached its peak in 1955 and mileage its peak in the following year. Separate figures for fleets of up to twenty-four vehicles (which were mainly in rural areas) and for those of over twenty-four vehicles show that the former carried their maximum number of passengers and ran their maximum mileage in 1951, significantly earlier. The Jack Committee regretted the lack of reliable statistics and concluded that 'the decline in the number of passengers carried on stage services by small operators seems to have begun as early as 1952 and to have continued at the rate of 3·4 per cent per annum',[1] while the rural routes of big operators probably also lost business earlier than did their main routes. It is likely that the decline began even sooner and that on the strictly rural parts of many routes, at least in Southern England, patronage was falling as early as 1949.

But to many people the problem of dwindling services in rural areas seemed to arise almost overnight. Little was heard until about 1956, but by 1958 the position had become so serious that a clear call was being made for Government action.[2]

[1] Rural Bus Services: Report of the Committee (Stationery Office, 1961).

[2] The first area from which came strong public complaints about reduced services was the rural North-East. A deputation met the Minister of Transport in 1956, after which the Northumberland Rural Community Council undertook a survey in mid-Northumberland to provide facts to reinforce further appeals. A full report was sent to the Minister and a booklet published in 1958. See *Northumberland Country Bus*, published by the Northumberland Rural Community Council.

## The Historical Background

This is not entirely surprising. While urban and inter-town traffic continued to increase, most operators were able to carry their rural losses without undue difficulty, and during the early 1950's costs per passenger seat were, allowing for inflation, steadily reduced by the use of bigger single-deck vehicles, more double-deckers, and more efficient maintenance practices. It so happened that many companies reached the maximum savings obtainable on the technical front just as traffic on most routes passed its peak. Rural losses then had to be treated far more seriously. Although further economies have of course been made, especially by the introduction of one-man services, after 1956 a bigger proportion of increased costs had to be passed on to the public, while the use of smaller cars was tending to keep static the cost of motoring.

This explains why so far as the general public were concerned the problem did seem sudden. But the onset of the decline must have been obvious had the trends been expertly studied. Unfortunately the industry itself undertook practically no research, and the subject interested neither the town and country planning world, nor rural economists and writers on country subjects. Valuable time for working out a long-term policy was lost. Even the manufacturer of the vehicles most popular among small operators failed to read the portents and ceased producing small (30-seater) buses, creating a shortage which considerably weakened a number of businesses.

Since 1958 the story of the bus industry has resembled that of the railways in rural areas. Had it been known that the Government would not help, at least for the next three years, curtailments would have been much more drastic, and many small companies in particular would be in a sounder financial position today. But operators were encouraged to let things slide in the belief that in the end justice would be done—that the Government would repeal fuel tax for public service vehicles or arrange some other subsidy in recognition of the fact that an extensive bus network was essential to the nation. The appointment of the Jack Committee in September 1959 and the publication of their recommendations early in 1961 naturally heightened their faith.

CHAPTER THREE

# RAILWAY MANAGEMENT

### WASTE IN THE WEST

DURING the eleven years 1950–61, British Railways closed completely or partially some 301 branch lines, or 19 per cent of the system. The estimated annual savings amounted to the equivalent of only 7 per cent of the railways' loss for 1960.

Far greater savings could have been achieved without closing a single branch line, although clearly many of the lines did need closing. Prodigious waste was rife throughout the system—a lack of foresight and imagination perhaps without equal in British history in a large industrial concern.

To be more specific, I should like to devote the first part of this chapter to the position as I have observed it closely in my part of the country—the South-West.

In twelve months in 1961–62 the Western Region produced eleven major economy plans for the South-West.[1] The total net annual saving forecast from the eleven was £248,643. A careful examination of the position led to the conclusion that over £1,000,000 yearly could have been saved on Western lines in this same area by piecemeal economies without closing any line completely, and without withdrawing a combination of trains carrying more than 20 per cent of any route's traffic.

This, of course, is not to say that none of the lines should have been closed; on the contrary, in the exercise of common sense, several of them should have been stopped many years earlier. But while the eleven cuts contributed relatively little toward

[1] The closure of the Helston, Perranporth, Launceston, Kingsbridge, Brixham, Exe Valley, Hemyock and Chard lines to passengers, some also to goods; the closure of the Ashburton branch to goods; the withdrawal of local trains on the main line between Castle Cary and Taunton; and the Taunton freight concentration scheme.

reducing British Railways' deficit, a much greater contribution could have been found with only a fraction of the inconvenience to the public—and incidentally with less damage to the railways' own interests.

Even had the £1,000,000 been saved and some of the branches closed completely, the West Country's railway system would still have run at a loss; the relationship between expenditure and revenue would, however, have been closer to that of more efficiently operated systems overseas. For example, Coras Iompair Eireann lost £2,000,000 annually on their 2,000-mile system mainly in countryside of scattered population.

The £1,000,000 could have been saved without violating agreements with the trade unions or disregarding statutory safety and other requirements, and it would not have involved spending large sums of capital. One-quarter of one year's savings would have sufficed. Timetables would have been recast, and diesel multiple unit trains in particular would have been deployed more economically, savings in rolling stock on some lines (notably the main line in South Devon) enabling the complete dieselization of more branches much earlier. A few individual stations would have been closed to passengers and a few to goods, while more stations would have become unstaffed halts. Some signal-boxes would have been closed, especially on single-track branches where a rationalized train service would have reduced 'crossing' requirements.

While making my inquiries, I found the staff at almost all levels active in pointing out missed economies. Anyone with a rudimentary knowledge of railway operation could see that no great effort was being made to keep to a budget. Often a full service with elaborate station and signalling arrangements continued long after the need for it had died; there was apparently no middle course between the full Victorian paraphernalia and complete closure.

A few examples follow.

*The Exe Valley branch.* In their submission to the South-Western Transport Users' Consultative Committee, the Western Region stated that by closing this line they hoped to save an annual £46,900—a gross £60,400 less the estimated loss of gross receipts of £13,500. The cost of major track and other renewals is not included. The submission claims that the use of single diesel cars working under the cheapest practicable conditions had been carefully considered. 'The most economical working arrangement would necessitate recasting and slightly reducing the present timetable and this would be likely to result in some loss of revenue. Even if the existing revenue were maintained, such a service would cost over £19,000 per annum to operate.'

Had such economies been introduced five years previously, the loss of over £45,000 annually could have been exchanged for what is referred to mildly as 'some loss of revenue'. Allowing a generous £10,000 for station costs with a reduced service, the total working loss over five years would have been £80,000 instead of £230,000.

*Plymouth, Tavistock and Launceston.* This is an example of a heavy loss being allowed to continue largely as the result of uncertain Government policy. Before 1962 British Railways were generally not able to close a branch line unless satisfactory alternative bus services existed or could be provided—if necessary at the railways' expense. It was found impossible to provide a bus replacement at an acceptable cost to serve Clearbrook and several of the intermediate points between Tavistock and Launceston. This, together with disagreement with the Southern Region about the future handling of milk traffic at Lifton and freight at Tavistock South, resulted in the entire route being kept open to passengers something like five years longer than the local railway management wished.

The closure plan at last went ahead in 1962, the obligation to provide a bus alternative having been removed. In their eventual submission to the T.U.C.C., British Railways stated that by closing this line—the whole of it to passengers, and most to all traffic—they hoped to save a net £61,466 annually. The gross working expenses were £72,000, of which only one-seventh (£10,500) was covered by gross receipts. The provision of a diesel railcar service 'coupled with simplified working methods' would involve 'movement costs' (excluding station costs) of £15,260 per annum.

Such a service would not pay, and in this case the continuance of even a smaller loss was probably not justified; but during its last years the service was costing thousands of pounds more than was necessary.

If movement costs for a diesel service were estimated at £15,260, clearly the line could have been run (inclusive of station and all routine costs) for no more than half the £72,000 gross which was being spent. That makes an expenditure of £36,000, from which we will deduct a gross revenue of £8,000, allowing a drop of £2,000 to offset the possible loss of traffic with a slightly reduced train service. The annual loss would thus be £28,000. £28,000 instead of £61,000 over five years would have resulted in a total saving of £165,000.

*The Helston branch.* Unlike the first two lines, this was closed only to passenger traffic and remains open for goods and perishables. British Railways said they hoped to save £9,077 annually. The gross savings were put at £20,842, while the loss of gross revenue was estimated at £9,965 and £1,800 had to be allowed for the cost of extra road journeys to replace the trains.

In their submission to the T.U.C.C. the Western Region put the 'movement costs' of a service of diesel units 'operated under the cheapest working methods possible' at £16,700. As already mentioned, the comparable figures for the Exe Valley and Launceston branches were £19,000 and £15,260—these lines being twenty-four and three-quarter miles and thirty-four and three-quarter miles compared with the Helston branch's modest eight and three-quarter miles. The figures are hard to understand. In the last published timetable the weekly passenger mileage on the Helston branch was less than one-third of that on the Launceston line.

Assuming that British Railways' figures are correct, however, the cost of operating the line could on their reckoning have been reduced by £4,000, which means that instead of saving a net £9,000 annually by the closure, the Western Region would have saved only £5,000. The difference here is not so great, because the Helston branch had heavier traffic and also was run more efficiently than the other two lines quoted. In fact it came so near to paying its way that it might be thought—bearing in mind that it was to be retained for freight and perishables anyway—that calculations could have been taken a step further.

In effect British Railways closed the branch because they

## Waste in the West

said they could not afford a £5,000 working loss, that figure being based on a service operated by diesel multiple units. But how many trains? The last timetable showed eight daily trains in one direction and seven in the other, two trains sometimes being required simultaneously. The high figure of £16,700 alleged to be necessary to pay for a diesel unit service suggests that exactly the same timetable was being used as basis. To arrange for one diesel car to cover all journeys would, however, have been relatively simple. Moreover, here was a classic example of a route on which the majority of traffic was carried by a small number of trains. Four well-timed trains in each direction daily would have covered most travel requirements.

Allowing for a trip from Penzance to Gwinear Road and back, such a service would have run about 100 miles each day, and at 6s. a mile—the accepted cost of running a diesel unit—the working costs would have come to something like £9,000 annually, or £1,000 less than the estimated receipts. The smaller service would have led to some reduction in receipts, but not a great deal, and thus the operating loss might be estimated at between £1,000 and £2,000 a year. To this must be added the difference between the cost of track maintenance for a freight service only and for passenger trains.

It could be added that in view of the heavy freight and perishables traffic, diesel locomotives and ordinary coaching stock would have been more convenient—precisely what in fact British Railways did use shortly before the closure. A skeleton service for passengers, sharing the motive power employed on the freight side, would have met the needs of commuters and long-distance travellers by one or two of the most popular trains, including the Cornish Riviera Express. Mileage would have been cut by half except on summer Saturdays, when several hundred people used the line in each direction and a fuller service would have been economic.

There would have been a loss, but had British Railways presented figures on this basis, would the T.U.C.C. and the Minister of Transport have agreed to the closure? The inconvenience in this case has been considerable, and the Cornwall Education Authority, the Royal Navy (8,000 travel warrants were issued each year), the Post Office, the holiday industry and the general public have had without question to increase their transport expenditure by a far greater sum than British Railways need have lost by keeping the line open.

### THE BURDEN OF THE PAST

Since the war more people have offered advice on how to run branch lines than on any other subject of comparable commercial importance. Railway enthusiasts, passengers, would-be passengers, and even motorists who never use trains seem to think they have the key to the railways' rejuvenation and have produced a welter of contradictory suggestions.

Most of the clamour has arisen out of self-interest, and most has had to be ignored; but with a few outstanding exceptions, the impression has been given that not even the saner suggestions from responsible organizations and individuals have been taken seriously. Not only have the public's views been largely brushed aside, but until 1962 British Railways undertook practically no fact-finding themselves to establish the true position of rural services.

It would be possible to formulate broad principles for the efficient running of a branch line as for any other business, but little more is understood of the subject now than was known in the 1850's and 1860's. Incredibly, in 1959 I was the first person to discover how the closing of a branch affected the travel habits of its former users. Despite repeated arguments about the amount of inconvenience caused, neither British Railways nor the opponents of their closure policy had seen fit to find the facts even in a specimen case.

An impressive point is that while expert advice is readily available to any farmer, and is often offered even when not positively requested, for decades no senior officials—nobody paid a salary comparable to that of Agricultural Advisory Service officers—has shown more than the most cursory interest in the well-being of many country stations. (Taking an annual trip in a non-stopping inspection saloon whose timetable is known in advance does not, of course, reveal very much.) While productivity has leapt forward on the farms, work goes on at the stations much as it did in Victorian times. The com-

## The Burden of the Past

parison might be taken a step further. The N.A.A.S. employs specialist economists and the study of specimen accounts has become a fine art, while the work of the officers in the countryside is backed by projects at universities and research establishments. The transport equivalents have been almost wholly confined to the engineering side; no work at all had been done before 1960 on the economics of railway working in rural areas, except where specific possible closures were being investigated. The Ministry of Transport had no expert advice on which to call; it undertook no fact-finding itself.

Had a commercial firm known as little about a substantial part of its activities as British Railways knew about their secondary and branch lines until 1962, bankruptcy would have been inevitable. In the first two chapters I touched upon some causes of the malaise: Government policy was uncertain, the railways lacked leaders with enthusiasm and foresight, and though large sums of capital have been poured into the industry, money has been short for urgent schemes.

It is easy to curse the politicians but, again, the Government line might have been different—or at least more decisive—had top railwaymen really demanded action. The roots of the trouble probably go back to poor selection of staff between the wars. It is no accident that many of the more efficient officers in high positions during the 1950's began their careers with the North Eastern Railway, the only pre-1923 company fully to realize the value of fresh blood.

While industry has increasingly appointed specialists and encouraged graduates by allowing them to start several rungs up the ladder, until recently nearly all railwaymen were recruited at the lowest school-leaving level. Jobs which industry would automatically allot to specially-trained experts have often gone to railwaymen in general offices with no qualifications. Ability has seldom been amply rewarded, length of service—coupled it is said with physical stature!—being the key to promotion. Lack of talent at the top and frustration among men lower down has led to many of the best staff leaving for other jobs.

It is thus that we find chief district officials bogged down with the real and imaginary difficulties which surround them. This is why stock reasons against making changes are trotted out

## Railway Management

before a proposal is properly considered—why a proposal has to be rejected lest the public should construe it as a precedent.

Failure to be alert and interested has often led to poor relations between staff at district headquarters and those outside. In many areas the gulf is absolute. Station-masters have long ceased passing even minor suggestions to headquarters, which in return do not bother to consult them on timetable changes.

Experience has not always been pooled even between adjacent districts, let alone between regions. The regions have gone their separate ways. Thus the London Midland either keep a station fully staffed or close it completely, refusing to introduce the unstaffed halts common elsewhere, while at one time the Western Region spent up to five times as much as the Eastern Region on annual maintenance of branches closed to passengers and used only by light freight traffic. Regional jealousies have often prevented the cheapest and smoothest working of traffic in border areas, and regional 'pride' has been costly in other ways. For example, when the Southern took over the Western lines in the Weymouth area, they immediately changed the quite modern Western signals for Southern ones—although most of the engine drivers over the route continued to be Western men.

But perhaps the worst waste of all has stemmed from lack of co-ordination between engineering and commercial departments covering the same territory. During the 1950's several branch lines were extensively relaid or resignalled shortly before closure. At one station—Clifton Mill on the London Midland Region—the office was actually being enlarged to take a new stove, which had just arrived, two days before total closure. Many slightly less extreme examples could be quoted—though probably not in the North Eastern and Scottish Regions.

Several examples have already been given of unnecessary waste allowed to persist during the very period that British Railways' deficit assumed alarming proportions, and more detailed examples are quoted in the second part of the book. The rest of this chapter is therefore concerned with points of branch-line management in more general terms.

## DIESELIZATION

One of the biggest blunders was to continue the mass-production of steam locomotives after the war. Over 1,000 were built between 1945 and 1960, many of them for local-train work. When British Railways' modernization plan was published in 1955, there were only nine main-line diesel locomotives and 179 multiple units (coaches with their own motive power) on the system. Thereafter diesel locomotives and multiple units were put on the rails as rapidly as workshops could produce them.

Not only should dieselization have been started earlier, but more of the initial effort should have gone to reducing costs on branch lines. Diesel railcars and multiple units for local traffic show a much greater saving over steam power than do diesel locomotives on long-distance express and freight services. On some branch lines the units reduced running costs from about £1 to about 5s. a mile. In addition to lower fuel costs, the units can be started at short notice, and be switched off when not required, no fireman or second driver is required for local trains, and the driving can be done from either end, saving time at termini and permitting easy running of through trains from branches to main-line destinations even when a reversal is involved. Several units can be joined to form a single train. Fuelling and servicing take much less time than with steam.

Steam enthusiasts point out with some justification that British Railways were apt to exaggerate the savings, for steam power was often used wastefully. If a two-car diesel carried traffic without difficulty, a six-coach steam train had obviously not been needed. But against this, even today the units are not being used to the best possible advantage because many timetables are still conceived in terms of steam power and too many branches are run as watertight compartments.

The public like the units, and on many routes they brought brisker traffic. But railwaymen generally placed too great faith in them, in some areas falling into the temptation of employing them on lines which had been almost totally forsaken and where the possibility of a resurrection of traffic was remote. On other routes (as in the Bristol area) the service was lavishly increased, especially in off-peak periods, with little prospect of a corresponding rise in takings. Inevitably there had to be experiments and mistakes, but the proportion of waste was unjustified.

The trains themselves are still often unnecessarily elaborate: single cars would sometimes suffice where two-car sets are used, and two-car sets instead of three. First-class seating should be omitted from more local trains.

### TRAIN SERVICES

Basically there are two kinds of service: that operating during most of the day for anyone on any kind of journey over the route; and that provided specifically to carry people on certain particular types of journey. Much of the waste in recent years has been the result of an unhappy compromise between the two.

Where the railway is still popular, it is usually desirable to run a well-balanced service. Passengers expect to find a train at roughly whatever time they wish to travel, and to avert long gaps it may be necessary to include a few trains which carry only light loads. The number of daily trips required for this kind of service obviously depends on the district: the more prosperous the population and the keener the road competition, the more generous the service which British Railways must run to hold their own. Except in remoter Scotland, five or six well-spaced daily trains are generally the minimum needed to provide adequate choice, and in many more populated areas the minimum is between nine trains and a train each hour.

Relatively few country services now come into this category. Increasingly, country people go by road for miscellaneous journeys, though they may still use trains for particular purposes, such as the daily trip to work, catching long-distance connections, visiting the seaside, or in some districts shopping, usually at well-defined hours.[1]

---

[1] 'Miscellaneous' would include travel for most social, religious, entertainment and medical purposes. Few country people attending any kind of organized event travel by train, although visits to hospitals may remain important where the railway still supplies the only public transport.

## Train Services

Clearly there is a point at which the railways should cut their losses and concentrate on serving only what might be termed the 'specialist' traffics. Rarely, however, have officials come to precise decisions. When receipts have dwindled, a service of say ten daily trains has been thinned to eight, and then perhaps to six, though this compares even less favourably with the hourly bus linking the same places. Frequently the individual trains have been removed without any effort at revising the service as a whole.[1]

In 1950, 80 per cent of the passenger traffic on many lines could probably have been retained by a third of the number of trains. Gradually some of the least-used services have been weeded out, but staff still outnumber passengers on many byways at slack times. Even where the public have long abandoned the railway for their general travel, the belief may persist that the timetable should 'look presentable'.

Of course the higher the daily mileage, the lower the cost per mile. But this is not the whole story. The little-used trips which merely 'fill out' the service and keep rolling stock fully employed are expensive if they add to signalling costs; on some lines signalling could be simplified if the freight service were given complete occupation for a couple of hours in mid-morning on Mondays to Fridays. Moreover, running the maximum mileage often hinders correct timing of the most important journeys. One train leaving a market town at 5.40 p.m., soon after the majority of shops and offices close, may well carry more passengers than a 5.20 p.m. and a 6.10 p.m. combined.[2]

Railwaymen sometimes argue that to keep a line open for merely two or three daily trains is uneconomic. Most branches inevitably incur a loss, but if 150 passengers can be attracted to three trains, why augment the loss by carrying only 30 more on double the service? There are indeed lines on which even a single train each way daily, serving commuters and giving a

---

[1] A most glaring example of this was furnished in June 1958, when the Western Region headquarters ordered a 10 per cent cut in most local services actually during the currency of the summer timetable. Most routes lost at least one daily train, but virtually no adjustments were made to the remaining trains.

[2] See the Moretonhampstead 'case history', page 135.

## Railway Management

long-distance connection, would make a healthy difference to local life. A diesel locomotive (suitable also for freight work) would be used instead of a multiple unit.

One other point on train services: connections between branches and main lines need to be tightened. Most people are not particularly concerned whether they spend four hours or four hours twenty minutes in total on a journey, but dislike having to kill the extra twenty minutes at a junction. Slack connections are usually an insurance policy: if one train is late, the passenger stands a better chance of reaching his destination on time. Generally he would prefer British Railways to allow him to take the risk. Many officials do not realize how non-railwaymen value their time. Answering a complaint about poor train-bus connections, the Western Region solemnly published the statement: 'It cannot be said that the present connections are hopeless. Margins of thirty to forty minutes should not be considered unreasonable.'[1]

### SIGNALLING

On many lines only half-hearted efforts have been made to reduce the signalling bill, which often exceeds the total gross revenue. Except where expensive electronic systems are installed, a signal-box has to be provided at every passing loop (or 'crossing station') on single-track routes. A signal-box open during the working hours of two men six days a week costs about £1,300 to £1,500 a year, making a fairly nominal allowance for upkeep of equipment. Obviously the fewer the crossing loops, the lower the bill—although other factors, such as level crossings, sometimes affect the picture.

Slight timetable reductions or adjustments would permit the reduction of crossing loops on a large proportion of branches. Often, indeed, full signalling could be abolished, and a branch (or on some longer lines perhaps the extreme end only) worked on the 'one engine in steam' principle. As already stated, sometimes the only 'crossing' movement is between the daily goods train and a little-used passenger train at a slack time. In effect,

[1] The Transport Commission's written reply to objectors to the proposed closure of the Exe Valley line, Devon, September 1962.

*General Administration*

British Railways frequently spend 5s. to 10s. per head to take passengers past a single signal-box.

So far from seeking to fit the train service to a cheaper signalling system, officials are often reluctant to take possible economies. Loops no longer needed for the regular service are sometimes retained (at £1,500 a year) so that delays can be slightly reduced on the rare occasions that extra trains may be run. Shortage of labour and materials is blamed for the delay in removing crossing loops and closing signal-boxes even where rationalization has been agreed upon. Had £500 been spent five years ago, often the saving by now would have exceeded £5,000.[1]

Level crossing control is costly. In some places full signal-boxes are retained where the job could be done more cheaply by a gate-keeper, but substantial economies can only be made where British Railways use automatically-controlled gates or seek Light Railway powers and leave level crossings unprotected. There are real difficulties here, but the matter has not been tackled with the urgency that the financial situation demands.

### GENERAL ADMINISTRATION

There is virtually no administrative system for most branches. Individual officers at district headquarters consult (or do not consult) individual station-masters along the route. Nearly all of the dozens of station-masters I have interviewed during survey work have criticized the lack of contact between themselves and district offices. It should be added that many station-masters are merely dignified porters—occasionally the station-master is now the sole employee—and not necessarily men one might readily expect to exercise authority and judgment.[2]

The most useful single move would be steadily to abolish individual station-masters and to appoint a linemaster responsible

---

[1] See the Teign Valley, Moretonhampstead and Keswick line 'case histories'.

[2] The most frequent remark made by station-masters was: 'I've given up worrying,' usually coupled with a statement of the number of years left before retirement.

for the day-to-day administration of an entire line or run of intermediate stations on the main line. The advantages would be numerous. While station-masters at adjoining stations may regard each other as rivals, a linemaster would see his line as an entity and would be consulted by district headquarters on matters concerning it. Total staff costs would be reduced, only porters being employed at most stations; but the linemaster would rank higher than individual station-masters, and a higher grade of man could be retained on the job than is now possible in the countryside. He would be given a car for visiting his various stations, would also regularly visit district headquarters, and would attend conferences with other linemasters. (Already some station-masters control a number of small stations, but this proposal introduces a new principle.)

A linemaster, sometimes speaking directly to the district superintendent, would be able forcibly to call attention to weak points in the timetable, and to resist changes which would upset regular travellers. He would know just who these travellers were, and might be able better than other railwaymen to suggest a compromise to satisfy them and meet the requirements of district headquarters for a new junction connection time, for example. The best London connection would be less likely to be missed narrowly, year by year; stations would be more likely to display train departure times. Excursion facilities could be brought closer to the demand. The advice of a linemaster might prevent unnecessary expenditure on renewals and repairs.

More stations could be reduced to unstaffed halts, tickets being issued on the trains. (A little adaptability could overcome possible hitches: for instance, the linemaster might travel on the first up train on Monday mornings to help issue the weekly season tickets.) Station buildings need simplifying: in some cases to erect a small new building would be cheaper than to keep up the existing range. Cheap fare facilities need rationalization: printing and other costs could often be cut if a single fare were quoted from a group of stations instead of individual fares varying by an odd copper or two. In some cases country stations might stock the junction's excursion tickets, of course also issuing passengers with a ticket to the junction: 'rebooking'

*Conclusion*

at junctions to take advantage of their better range of cheap tickets artificially reduces the receipts of some branches.

## CONCLUSION

It is far harder than realized by most people opposing branch-line closures to bring traffic back to the railways once it has been lost. The possibility of raising the revenue of many lines sufficiently to cover the cost of one signal-box is remote. Efforts should of course be made to attract business. Improved connections with main-line trains, and more through services from branch lines to main-line and other branch-line destinations can help, especially where the public have some assurance that the changes are permanent. Better publicity and attractive fares, especially on lines in tourist districts, can play their part. But the reduction of expenditure offers infinitely greater scope. It is in failing to prevent needless waste that British Railways have sorely jeopardized their future.

CHAPTER FOUR

# WHEN A BRANCH LINE IS CLOSED

### THE UNNECESSARY LOSS OF PUBLIC GOODWILL

M.P.s and correspondents in the Press often stress that in the matter of branch-line closures not only should justice be done, but it should be seen to be done. Yet in many cases there has been serious cause for complaint at almost every stage of the procedure. Closing branch lines is bound to be an unpopular procedure, but with the exercise of elementary foresight and tact much of the strain on public relations could have been avoided.

The regions have different ways of deciding that the time has come to close a particular branch line. Some closures are planned carefully, while others are hurried through when the region's over-all loss urgently needs reducing. With few exceptions, however, plans are produced piecemeal, neighbouring branch lines often being the subject of separate memoranda to the Transport Users' Consultative Committee (T.U.C.C.) and of separate inquiries. The railways and the public would benefit if intentions for the whole of one district were discussed together.

Another complaint is that British Railways often will not admit that they seek to close a line even when the decision has been taken by regional headquarters. A line is always stated to be 'under investigation' until the formal submission is made to the T.U.C.C., perhaps months later.

Even then there is no formal Press or other statement. Rarely are the Press sent a copy of the submission. A closure plan is thus frequently announced in the local paper following the first meeting of the council or some other organization which has received a copy; how much detail the reporter may be able to glean usually depends on whether or not the clerk of the meeting allows him to copy from the document. Otherwise the only intimation of the closure plan is given by notices displayed at railway stations. These notices are in formal language and small type, and alluringly headed 'Public Notice'.

One of the commonest criticisms is that British Railways not only fail to try to augment traffic on a branch line but sometimes positively remove the best services so as to lower receipts before the formal closure plan is presented to the T.U.C.C. Allegations that the traffic has been 'deliberately run down' have come from responsible people, and I have heard it admitted by senior railway officials. The public have no means of self-protection. British Railways can technically withdraw all but one daily train without reference to the T.U.C.C. (and in at least two cases have done so). Passengers upset by the cessation of certain trains can present evidence of actual hardship, but are not allowed to present evidence on apprehension. As most people have to make alternative arrangements the moment their train ceases, and as any inquiry by the T.U.C.C. would be likely to take months rather than weeks, it is hardly surprising that few complaints are formally presented.

Withdrawing the best trains in advance of complete closure may sometimes be a method of 'jumping the gun'. Thus an official explains: 'We are going to close this line anyway, and before we do so it is only sensible to reroute the through traffic.' But in theory no closure is settled until after the proper inquiry. Another form of anticipating events is the removal of branch lines from timetables and 'runabout' tickets before the result of the inquiry has been announced. Justifying this, officials have sometimes argued that they had no intention of jumping the gun, but from previous experience they were fairly certain that the closure plan would be accepted by a certain date by the T.U.C.C. and the Minister of Transport.[1]

[1] When planning to close a branch line, the Eastern Region wisely add a note in their timetable to the effect that the service might be withdrawn during the timetable's currency. In the past some other regions have at times omitted the tables of branch-line trains in advance of the closure plan being approved; if approval has not been received by press date, they now include the tables but without warning of possible closure, even when this is likely to take place midway in the timetable period.

*When a Branch Line is Closed*

Perhaps even worse than the railways' own shortcomings in public relations has been the inability of the Transport Users' Consultative Committees to win the confidence of the public they represent. Before proceeding further, it is necessary briefly to outline the work and composition of these committees.

### THE TRANSPORT USERS' CONSULTATIVE COMMITTEES

The T.U.C.C.s, the first of which was established in 1948, correspond roughly with the consumers' councils of other nationalized boards. There are eleven regional committees and a Central Committee in London. Members are appointed by the Minister of Transport. Most of those sitting on the regional committees represent specific interests, such as agriculture and local government, but the Minister may appoint further independent members.

The committees have two functions: to make recommendations generally about the standard of train and other services provided by nationalized transport undertakings; and to investigate objections to plans to withdraw train services. The public chiefly knows them through the latter function, which has indeed absorbed an increasing proportion of their time.

Until 1962, the nine English area committees passed on all their recommendations to the Central Committee, which reviewed them and sent them on—with or without amendment—to the Minister. The Scottish and Welsh committees put their recommendations direct to the Minister. Until 1961 all recommendations by the Central and the Scottish and Welsh committees on branch-line closure plans were agreed upon by the Transport Commission: the Minister had no need to use his power to direct the Commission to implement them. Though the committees had only advisory power, in effect they decided whether or not British Railways should be permitted to close a branch.[1] Then, in 1961, the Minister reversed the recommend-

---

[1] Rarely, however, did the committees recommend continuance of services. Had they done so more often, it is possible that the Commission would not automatically have implemented recommendations. Conversely it has been argued that the committees agreed to most closure plans to ensure that their word would remain the final one.

## Where the Committees Have Failed

ation that the Westerham branch in Kent should remain open.[1]

The arrangements governing the T.U.C.C.s were altered by the Transport Act, 1962, in advance of the abolition of the Transport Commission and the vesting of the railways in the Railway Board. Under the new arrangements, though the English committees continue to route recommendations on general matters through the Central Committee, all recommendations on branch-line closure plans go direct to the Minister. But their scope has been limited. The committees are now concerned only with proposals entirely to end passenger services, and have powers to make recommendations only on the grounds of hardship. They cannot, for example, recommend that a branch should remain open at even a small loss to avert inconvenience; nor can they suggest that a branch could be run more cheaply.[2]

As already stated, until 1961 all the T.U.C.C.s' recommendations were implemented by the Transport Commission, and the Minister did not actively enter the process of inquiry. Now the Ministry undertake their own assessment in every case, even when no recommendation about hardship is received from the T.U.C.C.s. All power is vested in the Minister.[3]

### WHERE THE COMMITTEES HAVE FAILED

The failure of the committees has been adequately to identify themselves with the public they represent. Their awkward

[1] See the Central T.U.C.C. *Annual Report* for 1961. It discusses the important question of what is hardship and what is merely inconvenience. The Committee concluded that the withdrawal of trains taking 167 commuters daily to London amounted to hardship; the Minister classified this as inconvenience only.

[2] Another change is that the committees now confine their deliberations to the activities of the Railway, London Transport, and Dock and Waterways Boards. Previously the nationalized bus companies had also come under their review, though few recommendations were passed on bus matters.

[3] This is necessarily a broad outline of procedures. For a more detailed description see the revised *Handbook* of the Transport Users' Consultative Committees, published in 1962 by the Central Committee at 22 Palace Chambers, Bridge Street, London S.W.1. Also the Transport Act, 1962.

names, their complicated structure, and the apparent ambiguity of their status does not help.[1] Fourteen years after the formation of the first of them, their role is still not understood—even by many leading officials in local government.

A serious drawback is that though they should be regarded as the travellers' own representatives, they are financed and staffed by the transport industry and usually meet on railway property. The staff are, of course, answerable only to the committees, but it would be scarcely surprising if they had greater sympathy with railway officials than with the people they represent but with whom they are less closely in touch. Whether or not the railway background in which the committees work exerts any influence, it obviously weakens their independence in the public's eyes.

The difficulty is accentuated by the fact that the members of the committees are public-spirited individuals by no means necessarily with specialist knowledge or even interest in transport. They might in fact use only private cars. There are thus no independent experts to challenge the railways' statements.[2]

On occasions the committees have exercised their independence and keenly criticized railway management, and in particular the facts and figures presented in support of closure schemes. An outstanding example was the condemnation of the manner in which data were presented to support the 1957 plan to close the Lewes–East Grinstead line.[3]

Following this dispute, the Commission and the Central

---

[1] Thus they are bodies to whom the traveller who cannot obtain satisfaction from a transport operator 'may appeal and get an impartial hearing', but 'this appeal is not to a detached judicial authority but to an influential body of transport users'.

[2] Until the revision of arrangements in 1962, two representatives of nationalized transport undertakings actually sat on each committee, and at one time they even had votes—see the Coniston 'case history'. The representatives were of course drawn from the higher—and presumably more persuasive—ranks of management.

In this context it should also be noted that at inquiries only actual users of a service, or organizations representing actual users, have been allowed to make statements. This has denied a platform to transport economists and others most able to question technicalities.

[3] *Report on the Proposed Withdrawal of Train Services from the Lewes–East Grinstead Branch Railway* (Cmd. 360, 1958).

## Where the Committees Have Failed

T.U.C.C. agreed upon a standard form of presentation of data, and though this has severe disadvantages, on balance it has been an improvement.[1]

But reading through the reports of the Central Committee, it is hard to resist the conclusion that generally the committees have been more concerned to put the producer's point of view than truly to represent the user. Indeed, sometimes the sweet reason of the railways is directly contrasted with the unreason of the public. For example, the 1958 annual report recorded 'our appreciation of the ready manner in which the Commission accepted our proposals' (for the 'agreed' scheme of presenting data just described) and added: 'In spite of this, we are aware that some bodies and members of the public are still not satisfied with what has been agreed on their behalf, and wish to be given further figures involving quite unreasonable detail.'[2]

Before and after the standard method of presentation of data was introduced in 1958, it was made clear that arguments from the public over the Commission's estimated losses and savings did not impress the committees, which were chiefly concerned with 'the exposition of any undue hardship which may fall upon people who are deprived of a transport facility'. Albeit lacking independent expert advice, the committees held that they themselves 'sufficiently represent and guard the public interest' and thus 'there is no need to provide information for anyone else, particularly regarding figures of savings, costs or losses, which can only be explained with difficulty to those unfamiliar with them, and can be distorted by facile and unsound reasoning'.[3] Some committees refused to allow objectors to

[1] See 'The Withdrawal of Railway Services' by M. Howe and G. Mills, University of Sheffield, in *The Economic Journal*, June 1960, and 'On Planning Railway Investment' by the same authors in *Bulletin* of the Oxford University Institute of Statistics, Vol. XXIV, No. 1 (1962).

[2] The same report included another dig at the public its authors were representing. It stated that objectors in one case presented their evidence sensibly and moderately—'by welcome contrast to some of the exaggerated attacks which have been made in the past against the capacity and honesty of the Transport Commission'. Doubtless some of the attacks were indeed unfounded and unfair, but the committee should have remembered that they were acting on behalf of the attackers, however unreasonable, and avoided generalized criticisms of this nature.

[3] Central T.U.C.C. *Annual Report* for 1957.

question the railways' figures, while cross-examination of railway representatives at public inquiries has always been held out of order. The new 1962 arrangements deprive the objector of all effective rights:

> 'Even though the committees and objectors may have before them a statement of losses or savings, the assessment of these is entirely a matter for the Board in the first place, and for the Minister when he decides whether to consent to the closure. There will therefore be no point in future in the objectors coming to Consultative Committees seeking to question these figures.'[1]

It must be conceded that some objectors, given the chance, would adapt figures to prove almost any point they wished. Clearly there must be some check on open discussion of finance and operating technicalities. But does that mean a complete embargo? It is worth repeating that at least in theory downright hardship is now the only relevant consideration. Neither the public, nor the T.U.C.C.s which represent them, have the means of questioning whether a branch line could not be run more cheaply than British Railways declare, and whether the loss might be so reduced as to make retention of the service economically justifiable.[2]

If British Railways ran their rural services efficiently, these questions would not arise so forcibly. But as shown in the last chapter, waste and inefficiency are widespread and frequently admitted by station-masters and other staff. There does often seem sound cause to criticize the railways' estimate of savings, calculated after all by individual officers. It is easy to understand the anger of the informed objector who is forbidden to point out that his community could be served at a loss insignificant compared with that being incurred at the time the

[1] The revised *Handbook*.

[2] As an example, I would hold that retention of the Helston service (see page 24) was economically justifiable, its value to the public exceeding the small sum necessary to subsidize it. The extra transport costs borne by the Services, the Post Office, the education authority and other organizations and individuals alone exceed the subsidy that would be required, while much inconvenience—if not actual hardship—could have been avoided. Nobody questioned British Railways' estimate of the cost of running a simplified service, though compared with the estimate for running a similar service on two other lines, the figure was unrealistically high.

## Where the Committees Have Failed

closure plan was formulated. Yet if he approaches the railways or the Ministry of Transport he will most likely be told that the proper process of consultation has taken place. It is, indeed, possible that transport users would be in a stronger position if the railways and the Minister had to answer complaints themselves.[1]

When the consultative machinery was designed, prior to nationalization of the railways, the Minister of Transport promised that the committees would for the first time give users 'a real means of making their views felt'. But in practice, as already said, the T.U.C.C.s have more often been concerned to put the railways' point of view to the travellers. Their desire to moderate criticism and not to add to the management's embarrassments is natural, even though it may not have been in the best long-term interests. The real failure of the committees is that they have contributed little towards resolving the basic differences.

The 'consultative' committees were designed to advise on policy making, and the Transport Commission had a statutory right to consult them and take their views into account on matters of general policy. The fact that even the Central Committee have largely kept to routine matters is their own choice. They have remained silent even where they would seem called upon to give a lead.

Well before 1961, when the recommendations on branch-line closures were accepted automatically, it was clear that they were being expected to perform the impossible. Already British Railways were supposed to be paying their way, yet the Committee had to consider whether the social hardship might outweigh the economic cost, and were prepared to say that some lines should remain open at considerable loss to be borne by the railways, and in other cases that the railways should subsidize the replacement bus service.

To quote two independent observers of the work of the Committee:

---

[1] This point is also made by G. Mills and H. Howe in 'Consumer Representation and the Withdrawal of Railway Services' in the Autumn 1960 issue of *Public Administration*.

## When a Branch Line is Closed

'The Central T.U.C.C. could have made some attempt to resolve the contradiction by putting forward some policy of its own to represent the views of the users. Detailed policy could have followed one of three general courses: (*a*) withdraw most, or all, unremunerative services; (*b*) subsidize specific services; (*c*) regard the railways as a whole as a public service, and grant a general subsidy.'[1]

Likewise the Committee could have pointed out forcibly that more branch lines could have been closed if better connections were made between long-distance trains and buses than were generally found where branch trains had already been withdrawn. (See page 52.)

[1] G. Mills and H. Howe, *Public Administration*, ibid.

CHAPTER FIVE

# BUS SERVICES

### AN OVER-CONFIDENT INDUSTRY

By and large, British bus companies serve the public well. Buses are of course cheaper to run than trains, but the difference in cost is frequently accentuated by the greater efficiency displayed in road operation. In particular, bus companies carefully avoid unnecessary expenditure and watch overheads. Their housekeeping is sound. It would be hard to find bus equivalents to the more glaring examples of railway waste cited in Chapter 3.

But the bus industry has one great weakness, which already produces some undesirable results, and which could seriously impair future services as conditions become more competitive. It is too self-reliant. With inadequate fact-finding, poor public relations, and lack of readiness to follow each other's best operating practice, many companies are too sure that they know what best suits their own and the public's interest.

Critical self-appraisal is urgently needed, and the industry might well benefit from a greater degree of examination and supervision of parts of its activities by the Traffic Commissioners.

### THE PROVISION OF RURAL SERVICES

Before considering certain aspects in detail, however, a few essential factual points have to be made.

As already stated (Chapter 2), most of the large 'territorial' English and Welsh bus companies belong to the Tilling or the British Electric Traction groups. The State holds all the Tilling shares, and has a substantial interest (in many cases 48 to 49 per cent) in most B.E.T. companies.

Under the 1962 Transport Act, the State bus companies have been transferred to the Transport Holding Company, entirely independent of the railways. Previously, though nominally there had been close liaison between the bus companies and the railways, in practice co-operation was limited. The position of the Tilling companies was particularly anomalous. Although wholly owned by the railway-dominated Transport Commission, and including local railwaymen on their boards, they none the less competed freely against train services and frequently omitted to take the most obvious steps towards integration of road and rail schedules. Common ownership yielded few benefits, and to some extent restricted both the companies and the railways.

Although smaller operators still provide many essential services, especially in areas of sparse population which did not attract take-over bids and amalgamations in the past, the majority of rural services are now run by the territorial companies. Many are 'cross-subsidized' by more profitable urban and inter-urban routes. In 1962 the total loss of all 'unremunerative' rural routes was probably between £4,000,000 and £5,000,000.

A few regional examples may be quoted. In the area of the Lake District Transport Inquiry, the annual net loss in 1959–60 on all stage services was £85,000. The total loss made by all the uneconomic services was £120,000, made up as follows:

|  | £ |
|---|---|
| Ribble Motor Services Ltd. | 87,034 |
| Cumberland Motor Services Ltd. | 24,600 |
| Small operators (estimated) | 8,000 |

Some 6,000,000 miles were run annually by Ribble Motor Services in the survey area at an average loss of about 10 per cent, equivalent to roughly 1 per cent of the company's total turnover. Services in and out of Penrith alone lost £25,000 annually —equivalent to a 5s. rate if the whole burden had to be carried by the town itself.[1]

The Jack Committee reported that 64 per cent of the routes and 32 per cent of the mileage run by the Birmingham & Midland Motor Omnibus Company Ltd. was uneconomic,

[1] This is of course a fictitious example, for if losses were borne by local authorities they would be shared by the larger surrounding rural area; but it serves to emphasize the scale of 'social transport'. An independent operator also made heavy losses on services based on Penrith.

## The Provision of Rural Services

while the Tilling group put their 1958 total loss on unremunerative rural mileage at £1,950,000, and the Scottish Omnibus Group at £585,000.

The Highland Transport Inquiry reported that 25 out of 36 of MacBrayne's stage-service bus routes ran at a loss: 'It is clear that if part of the Company's administration costs and other overheads which pertain to bus services is taken into account then the MacBrayne bus undertaking is quite unremunerative.' The Highland Omnibus Company was also making a loss, stated the report of the Inquiry: 80 per cent of the routes outside the burgh of Inverness 'are on a strict accounting unremunerative'.[1]

The Council for Wales and Monmouthshire also conducted an inquiry into the rural transport problem and reported: 'The Council have been told that about a quarter of all the services operated by two of the largest companies in Wales can be described as unremunerative rural services and that, in the case of a third large company, more than 98 per cent of the annual mileage operated in one Welsh rural county is run at a loss.'[2]

But 'losses' and 'cross-subsidization' are not so simple as they might appear—and as they apparently appeared to the Jack Committee. The standard practice among bus companies is to express expenses and income in pence per mile, averaged over a complete district or substantial group of routes. This rough-and-ready costing produces results which would be acceptable to no economist in competitive industry.

The Public Transport Association told the Jack Committee: 'Many unremunerative rural services which can be operated today in combination with more remunerative routes would, in fact, disappear, if they had to be run independently, because their independent costs of operation would be prohibitive.' Two points should be made here.

Firstly, according to the P.T.A., a bus is unremunerative if it is not earning sufficient to cover the *average* expenses including overheads. But if a rural service is run entirely in 'layover' time,

[1] Highland Transport Inquiry: *Bus Services in the Highlands and Islands* (Stationery Office, 1961).
[2] The Council for Wales and Monmouthshire: *Report on the Rural Transport Problem in Wales* (Cmd. 1821, 1962).

wages become an overhead, and the cost may thus be reduced so much that the company would be worse off if they closed the route, even though it may be yielding revenue well below the average cost. A route is really only uneconomic if the operator would be positively better off without it. It has to be borne in mind, too, that traffic on the main routes would decline if all uneconomic feeders were abandoned. The closure of all uneconomic routes would mean that some now profitable would begin to show losses.

Secondly, though of course large companies could not operate rural routes in isolation, the statement that 'their independent costs of operation would be prohibitive' needs qualification. In many cases small independent operators would find them profitable. The fact that the territorial companies are unwilling to hand over such routes to small concerns, and indeed sometimes go to extreme lengths to sustain their monopoly position, suggests that much rural mileage is not quite the liability portrayed by the Public Transport Association.

In some instances, certainly, the Traffic Commissioners have granted exclusive licences for profitable routes specifically on the understanding that the operators shall bear some of the essential uneconomic mileage. No doubt the Commissioners would generally be more disposed to permit competition on profitable runs if the large operators ceased to treat their rural obligations seriously.

Against this, it is not widely realized that though an operator has to obtain the consent of the Traffic Commissioners to introduce a new route or to alter an existing one, a licence for a complete route may be given up without permission or notice. Moreover, though in theory the Transport Users' Consultative Committees were concerned with State bus services until 1962, in practice few bus matters were placed before them, and there was never any suggestion that the abandonment of a bus route should be subject to the kind of scrutiny made when British Railways wish to close a branch line. Technically bus companies can hand in their licences for all uneconomic routes at any time.

This suits many leading officials in the industry. The companies themselves are both judges and champions of the public's

## Self-knowledge and Public Relations

interest, deciding what rural losses shall be maintained and who shall pay for them.[1] Indeed, the only limit to their power is their possible reluctance to call attention to it by abruptly abandoning large sectors of the countryside.

### SELF-KNOWLEDGE AND PUBLIC RELATIONS

The strength of the bus industry is its simplicity of organization. But the administration of some companies is too streamlined, inadequate efforts being made to discover traffic and other trends, and public relations being sadly neglected.

As on the railways, too often a gulf yawns between staff on the job and officials at headquarters. Even among the largest companies there are some without senior executives who go out to examine problems and potentialities on the road. Much thus depends on the district inspectors, virtually the sole go-between.

Some inspectors are first-class men, respected equally by the drivers and conductors under them and by the management. But the very companies which take no other measures to bridge the gap between bus crew and office tend to appoint inspectors of limited experience and outlook. Certainly promotion should be encouraged, but automatically to appoint inspectors from the ranks of the drivers, coupled with the lack of thorough training schemes, is scarcely likely to supply the necessary administrative ability.

Generalization is dangerous, for practice varies so widely. At one extreme, for example, can be quoted Ribble Motor Services, which maintain a network of able inspectors and a statistical department which keeps headquarters accurately informed of traffic trends, while executives from head office frequently tour the company's area and welcome the opportunity to meet local councils.

At the other extreme are the companies whose inspectors feel as remote from management as do many British Railways' station-masters—companies which take all decisions without

---

[1] As examples on the broad scale, Ribble decide that South Lancashire shall subsidize the Lake District, and Crosville Motors that the Wirral Peninsula shall subsidize North Wales. South Lancashire and Wirral passengers of course pay artificially high fares.

consulting the public and whenever possible avoid meeting councils. Timetable changes generally stem not from positive, considered action but from the piecemeal need for economy or from public complaint. Indeed, it is often held that a service must be satisfactory if complaints are not received. When asked how he ensured that services in a particular district were being run efficiently, the traffic manager of one company told me: 'We have carefully investigated every complaint.' His conception of efficiency went no further.

Complaints and suggestions of course play their part; but again it is the forward-looking companies, which encourage public interest in their affairs, which are the most likely to be approached by passengers and would-be passengers. In some other cases the passengers' prevalent feeling is: 'It's no use complaining; they're not interested.' Letters are not always even acknowledged.

The public relations of some companies leave almost everything to be desired. Requests from the Press for information, for example, are often curtly refused or ignored. Few bus companies have Press officers readily available. Although there are exceptions, the Tilling companies have been particularly notorious for their evasiveness.

How not to treat local authorities may perhaps be demonstrated by giving two examples concerning Totnes Rural Council in Devon. Having numerous public transport matters on the agenda, the council persuaded bus and railway officials to meet them and neighbouring councils jointly. At the start, the bus spokesman said he would like to emphasize that his attendance was experimental and did not create a precedent—an announcement which hardly warmed the air. Then, the council sought to have a particular bus timed five minutes earlier to enable passengers to catch a train. Without in any way commenting on the proposal or agreeing to consider it, the bus company replied that if the council 'objected' to the timetable, the proper course was to do so through the Traffic Commissioners.

'Objection' and 'complaint' are ugly words. 'We think that many complaints made by the public about "inadequate" bus services would be withdrawn if the reasons for the situation were

## Self-knowledge and Public Relations

explained,' said the Jack Committee. 'Many criticisms would never be voiced at all if explanation preceded an action which might cause inconvenience.' Relations between operators and passengers 'can and should be improved', public dissatisfaction with operators and the operators' weariness of complaint producing 'an atmosphere inimical to the solution of many bus problems which could easily be solved given a little goodwill and tolerance on both sides'. The Committee made no detailed proposals but noted that the willingness of senior representatives of the companies 'freely to meet people, in small groups or in public, though it can make heavy demands on the time of busy men, seems to have been worthwhile'.

As traffic declines, it becomes more urgent for bus companies to keep closely in touch with their customers' needs and habits, and to lose no opportunity to improve the ratio of expenses to receipts. Not only do conventional public relations need re-planning in many cases, but active assessments of local feeling should be made and suggestions should be invited. For instance, a well-publicized meeting could be called to tell a village why its bus service will have to cease unless fares are increased. With adequate explanation, it is possible that a village might agree to support above-average fares to retain its link with the outside world. At least it would be worth trying—a point made by both the Jack Committee and the Council for Wales and Monmouthshire. Similarly, in certain circumstances utility buses might be offered as an alternative to complete shut-down.

Above all, from time to time each route or group of routes should be scrutinized in detail by a senior official. He could spend some days travelling on the buses, talking to passengers, examining connections and turnround times, and would call on the chairman of the parish council, the largest employers, and anyone else likely to vent useful ideas. Later in the book, an example is quoted of how a few days spent in a Devon valley might have rewarded the bus company.[1] The employment of an outside research assistant to the traffic manager at a salary of £1,200 a year would suffice for this kind of work in medium-sized bus companies. The financial returns might be handsome —not to mention the value in goodwill.

[1] See the Teign Valley 'case history', page 126.

*Bus Services*

### ROAD AND RAIL CO-OPERATION

Connections (or the lack of them) between bus and train services are probably the subject of more complaints and misunderstandings than any other aspect of the rural transport problem. Almost every country person who uses public transport—and for that matter many who do not—has a personal story concerning a narrowly-missed connection, or a suggestion for improving connections. The impression is given that the railways and bus companies positively delight in demonstrating their independence of each other. Probably most people believe that the one thing needed to bring about better connections is a change of heart by the operators.

A culpable lack of initiative is apparent in many individual cases, and undoubtedly does the industry direct and indirect harm. But it must be stressed that with the utmost goodwill and common sense, transfers from train to local bus will remain uncertain and frustrating. Conditions of course vary from place to place, but troubles will continue to arise wherever long-distance trains are repeatedly late, and wherever the connecting buses carry more local passengers (those travelling by bus only) than people completing lengthy journeys.

Road and rail connections were studied closely as part of the Lake District Transport Inquiry, and in the Lake District in particular it was found that the tasks of taking people home from shopping and from work, and of providing connections into the countryside from long-distance trains were just not compatible. Why, argued the bus companies, should twenty or more workers already aboard a bus at the end of a day be kept waiting half an hour for one or two passengers from a train running late?

Railway timekeeping in the Lake District was poor. The bus companies could have provided satisfactory connections only by allowing lavish recovery time; they pointed out with justice that they had been able to keep their fares low largely because buses were used intensively, and that idle half-hours inserted into schedules in case trains should be late would add heavily to the loss on many routes. Another difficulty especially evident

## Road and Rail Co-operation

in the Lake District was the frequency of changes in the railway timetable. Even if sent adequate notice, the bus companies would have had inordinate trouble in adjusting their tight rotas to keep pace with the changes.

Examples were found of a railway and bus route running parallel where the bus had nearly all the local traffic and the railway catered only for the specialist traffic, mainly people beginning or ending long journeys. The railway could not have been closed without considerable hardship. Indeed, without grossly impairing its own efficiency, the local bus service could have taken over little of the railway's business. Most people who had used the trains would have worked out their own arrangements or ceased to make their journeys at all.

Generally, only when a bus is run specifically in connection with main-line trains does it satisfactorily replace a rural branch line. Rail-road co-operation works well in Eire, where C.I.E. themselves operate the buses as well as the trains, timekeeping is good, and the connections are published in the timetable. Similar circumstances are required to bring about sensible collaboration in most of rural Britain.[1] In many cases adequate connections are probably possible only if British Railways have their own buses, preferably painted in railway livery even if on hire from the local bus company. Of course this would mean duplication of buses, but there is a precedent for this in the separate express and local buses already run even by the same company over many roads; and a second bus would be infinitely cheaper than maintaining a branch line.

The Lake District Transport Inquiry included a special study of the railway from Penrith through Keswick to Cockermouth. This is the subject of the 'case history' starting on page 150, but the report on the question of a special railway bus service to replace the trains may be noted here.

> 'There is no doubt that in existing circumstances Keswick would be extremely badly served from the point of view of long-distance connections were its branch line closed. On the other hand, were

[1] This section is confined to the strictly rural problem—to the route with services so infrequent that if the traveller misses one bus he will be hours late at his destination. Where buses run every half-hour or more often, the difficulty does not arise.

British Railways themselves able and willing to run a limited-stop bus service similar to the present train service the position could be entirely different. Such a bus would have its place in the railway timetable. Through tickets would be issued for long-distance journeys both to and from Keswick, where the railway station (or a replacement office in the town) would continue to deal with parcels inquiries and so on. The buses would carry parcels and heavy luggage, and this might of course involve the use of trailers, which would be a much better proposition for a specialist service like this than for ordinary stage buses. The cost of such a bus service replacing the Keswick line would probably be £25,000 to £30,000 a year, of which half to two-thirds might be recovered in fares.'[1]

In many cases even a single daily railway bus, running to a junction station, with through fares, and conveying heavy luggage, would bring a small town or village appreciably nearer the outside world. Country people may make few long-distance journeys, but when they occur they are occasions of great importance. Interviewing conducted in the three survey areas for this book and my other reports emphasized that anxiety about being cut off from the only truly national transport system is mounting.

At the time of writing, British Railways are beginning to hire buses to take through passengers to holiday places which have lost their train services. It is to be hoped that the principle will be widely adopted as more branches are closed and larger towns and tracts of country are removed from the railway system.

Widespread improvements in connections cannot be expected until there is a positive policy of co-ordination.[2] But the

---

[1] The branch line made a loss of about £50,000 in 1960.

[2] It is relevant to quote in full the paragraph on co-ordination of transport services in the Council for Wales and Monmouthshire's report on the rural transport problem:

'It seems to the Council that insufficient attention is being given to the possibility of arranging better co-ordination between the different parts of the public transport system in the rural areas so as to ensure that the best use is made of the limited resources now available. They are concerned that what appears to them to be a gap in present administrative arrangements should be bridged and have carefully considered whether any existing body might serve this purpose. It seems however that no official body at present carries a responsibility which extends to the day-to-day operation of both road and rail services. The Council therefore fact that even in some easy cases it is not thought necessary to provide them again emphasizes the self-confidence of the bus industry. The railways' failure to encourage road feeder services has been yet more glaring. The buses which British Railways subsidize as replacements for branch lines do not generally connect with trains, and in at least one case (Coniston) where connections are made they are not mentioned in the railway timetable. Only the Southern Region (in the West Country) and the Scottish Region (notably in the Highlands) reckon to give passengers the elementary information needed for completion of journeys to places off the railway system.

Where passengers are supposed to be able to transfer from trains to buses, the arrangements are often unimaginative. Direction signs are lacking at stations, shelter is not always available for waiting bus passengers, and heavy luggage is seldom accepted unless by special favour of the conductor. Often railway and bus staff have no form of liaison: a bus may wait for a predetermined number of minutes should a train be late, and then move off just as it pulls into the platform.

#### SMALL OPERATORS, TENDERING AND SUBSIDY

The large territorial bus companies are hotly opposed to a subsidy system based on an independent assessment of social transport requirements and on competitive tendering between operators to fulfil essential uneconomic services with the aid of subsidy. They themselves like to be both judges and champions of the public's interest, and seek an all-round abolition or reduction of the fuel tax, which would help them show generosity to rural areas according to their own principles. But their biggest objection is that a subsidy system of the kind envisaged by the

---

think that it would be to the general public advantage if a small independent standing committee—it need consist of no more than five or six persons—were set up to advise on positive steps which might be taken to achieve the better co-ordination and more effective use of existing public transport resources in the rural areas. Such a committee, able to take an over-all view of rural transport arrangements, could also serve a useful purpose by undertaking the preliminary examination of, and giving advice to the Traffic Commissioners on, applications for determination of the need for a rural bus service under the procedure previously described.'

## Bus Services

Jack Report, the Report on the Rural Transport Problem in Wales and the Highland Transport Inquiry would favour existing and new small-scale operators at the expense of the big companies.[1]

The number of bus operators in Britain has declined steadily, the big companies absorbing the smaller. Rarely does a big company sell a route, however uneconomic. Not only this, but big companies exert care that new operators shall not start or expand. (For example, when a small proprietor seeks authority for a route previously refused as uneconomic by the large company, the latter have second thoughts and themselves request the licence.) The dominance of the big companies does not therefore necessarily mean that smaller concerns would shy from playing a greater part were they given the opportunity. Indeed, there is considerable evidence that many existing small operators would be prepared to expand, and many newcomers to risk their capital—even should no subsidies be available—if the large companies ceded some rural routes. And if subsidies were granted on a tender basis, small concerns would most often put forward the better terms.

For a variety of reasons, small proprietors have lower running costs—often sixpence or ninepence cheaper per mile, including all overheads—than the nearest big company. Their standard of bus is generally lower. Old buses do not suit large companies, but suffice for many 'independents'. In this, as in much else, improvisation is the small man's prerogative. He and

---

[1] The Jack Committee recommends:

'Once the County Council had decided that a particular service was needed, its next step would be to publish details of what was required, together with an invitation to operators to tender for it. The successful tenderer would then apply to the Traffic Commissioners for a road service licence which presumably would be granted unless there were overwhelming reasons against it.

'The financial assistance in this case could be provided in a number of ways. One would involve an undertaking to make good the operator's losses; another would involve a fixed sum paid to the tenderer.... The second method would be simpler to operate and would have the additional advantage that the amount of financial assistance would be definite and not, as in the first case, an uncertain quantity. In fixing the contract, regard would doubtless be had to its reasonableness in relation to the services to be provided. The Traffic Commissioners might be asked to certify that the tender was in fact reasonable.'

## Small Operators, Tendering and Subsidy

his family may devote long hours to the work, he can make unorthodox arrangements such as carrying a commuter free in exchange for fare collection on the daily bus needing a conductor, or paying the driver only for his journey time to and from the cinema, not during the performance to which he is given a free ticket.

Often, of course, the buses are a sideline, combined with running a garage or other business. But even if a dozen or so buses are employed, personal supervision and goodwill are likely to be valuable. The proprietor knows what is happening in the village and can keep services attuned to demands. His staff is usually loyal and contented, mainly living in the village or suburb where the depot is situated. At slack times men will not be standing idle, for many prefer to work 'spreadovers', covering both morning and evening rushes, for small concerns, rather than complete their working day in one turn of duty for the large firm.

Managements of the combines tend to regard the competition of the independents as unfair because they evade many of the difficulties inherent in big business. In fact the small men probably merely exploit their own advantages in the same way that the combines exploit their size by bulk buying, modern servicing methods and mechanized accounting.

What is to be gained by the big companies retaining routes on which they consistently lose money when independents are ready to take them over and might run them profitably? Of course routes cannot be relinquished wholesale to newcomers with doubtful credentials, but steadily over the years the large companies could, if they wished, with the help of the Traffic Commissioners, appoint independents to run a proportion of rural services under their umbrella. Connections would be made where suitable, and the services would continue to figure in the big firm's timetable.

In particular the big companies should take advantage of the fact that many existing small operators, and garages who would probably be willing to run a part-time service, are situated in villages at the farthest end of the routes from the town central bus depot. In Devon there are a number of examples where the use of a private bus from one village to feed into a big company bus from another village could save separate vehicles being sent out the whole way from the town.

Unnecessary expenditure must be avoided, and if the combines are unwilling to make sensible arrangements to cut costs themselves, there is a strong case for the Traffic Commisioners or some other authority enforcing them. Most definitely the public would not benefit from the unplanned disintegration of the combines' rural networks—which might result if subsidy were introduced on the basis of the tendering system.

The disadvantages of the straightforward tendering system (the automatic granting of a service to the man offering the best terms) have been forcibly demonstrated by the case of school buses. Frequently school bus and car contracts (sometimes also hospital transport contracts) have been lost by men with uneconomic stage services to maintain to those concentrating entirely on the more profitable contract and private-hire business. Sometimes contracts have been lost to newcomers who have intimated that they will buy a bus should their tender be successful. Even on occasions when a local authority knows that such a tenderer may fail to provide a satisfactory service, questions are unlikely to be raised. Standing orders, political feeling, and anxiety to display how carefully the ratepayers' money is being spent, lead almost universally to the lowest tender being accepted.

The cheapest service by no means always gives best value in the long run. It has been widely recognized that without school contracts many operators would have to cease stage services—as some have already done. 'It has been made very clear to us that but for school contracts many stage service operators in the Highland area could not continue,' stated the Highland Inquiry. And when stage services end, the education authority is immediately faced with higher transport bills.

In 1957 the fares of school children travelling by stage buses in Mid-Northumberland totalled £920. If all the stage services ceased, the County Council estimated that the cost of chartering special vehicles would be £5,000.[1] In 1960–61, Westmorland County Council spent £12,100 on conveying 1,263 pupils

[1] Northumberland Rural Community Council: *Northumberland Country Bus* (1958).

## Small Operators, Tendering and Subsidy

by stage-service buses and trains and £20,100 on taking 1,175 pupils by specially-chartered buses, taxis and other vehicles. The Director of Education stated that if all stage buses and trains ceased, the cost of engaging special vehicles to take the 1,263 children to school 'would be approximately three times as much as what we are paying for bus and rail contracts: thus the £12,100 would be increased by some £24,500'. The report by the Council for Wales and Monmouthshire stated that in Merioneth the cost of transporting children on special buses was about twice as high as the fares on stage services, the cost by taxi being even higher.

Clearly education authorities have a vested interest in maintaining a substantial network of stage services, and the reports of both the Jack Committee and the Highland Transport Inquiry, as indeed the author's two previous reports, enjoined county councils to take account of the general stage bus services in their areas when awarding school contracts. In many cases, however, the county councils are inadequately informed of the possible effects of removing a contract from an established operator, and it is nobody's job to point them out. Certainly the Traffic Commissioners do not do so. And even if they did, it is unlikely that there would be a substantial deviation from the principle of placing the contract with the best bidder.

If the county councils cannot take broader issues into account when it is so blatantly in their interests to do so in the case of school buses, they are even more likely automatically to accept the lowest offer should a contracting system be applied to stage services on which a subsidy were provided.[1] Many unsatisfactory newcomers might be attracted to the industry, and present operators might withdraw rapidly from uneconomical rural activities. The position would be especially unhappy if tendering were on an annual basis, insisted upon by most councils for school bus contracts.

Small operators should be encouraged to play a bigger part

---

[1] The Jack Committee mention the employment of 'existing operators' as the county councils' agents, 'since it would use the existing resources and valuable experience of the bus industry'. The Committee did not, however, suggest that tenders should be restricted to established operators or that priority should be given to them over newcomers making better offers.

## Bus Services

in sparsely-inhabited areas. But especially where subsidy is concerned, this must be done in co-operation with the combines, under the auspices of the Traffic Commissioners or some equivalent organization. Local authorities lack the experience and the independence of judgment to administer subsidies. Here we are concerned with subsidies only so far as they affect the extremely important tendering system and the implications for the size of operators likely to run rural routes in future. Subsidies are considered generally in Chapter 7.

### THE TRAFFIC COMMISSIONERS

Each year the decisions of the Traffic Commissioners directly affect the travel habits of thousands of people and the profitability of the routes of numerous operators. It is, therefore, no small achievement that the Commissioners are probably awarded more compliments, and are the subject of fewer complaints, than any other part of the mechanism of public transport. They are respected equally by those who provide buses and those who use them.

It is because the Commissioners have such a clean record that I believe their powers and resources should be extended to administer subsidy and other matters. That is a question for discussion in Chapter 7. There are, however, ways in which the value of the Commissioners could be enhanced even without major reconstitution and these should be considered here.

Although judicial authority is exercised scrupulously, the Commissioners' background knowledge and understanding, and willingness to take initiative in offering advice or putting the public fully in the picture, varies greatly. Geography is sometimes an obstacle. To quote one anomaly, North Wales is in the North-West traffic area. A number of the areas are too large. Revision of boundaries and an attempt to induce the most self-effacing Commissioners to follow the example of the more enterprising might pay handsome dividends.

The Commissioners' independence as judicial tribunals must not be jeopardized. But they are so securely established that there would be little risk in their venturing to voice opinions more frequently, especially since there is no other authority

## Miscellaneous Points

to do so. A few words from them could smooth out many anomalies.

The Commissioners would be more effective if they exercised greater powers over chartered services including school buses. Many operators subsidize stage services out of private-hire and excursion work: here too the fat is used to fry the lean. Education authorities, for example, should either have to consult the Commissioners before accepting tenders for school buses, or the Commissioners should have to pass arrangements made by the authorities.

At present licences can be relinquished for complete routes without reference to the Commissioners. This should not be so. Application should be made as in the case of new routes, or variations to routes, and an opportunity given for hearing objections. The publication of the fact that the existing operator wished to terminate his services might lead to another operator seeking the licence and continuity being maintained. It is far harder to revive a service after a break.

Another useful change would be to give parish councils the right to appear before the Commissioners—as they do before most other tribunals and inquiries concerning matters of intimate concern to them. At present only county, borough and district councils have the right. Although a parish can of course be represented by its rural district council, its case may lose clarity by passing through an indirect channel. Parish representatives are sometimes heard, inasmuch as the Commissioners are entitled to hear whom they think fit, but a concession is no exchange for a right.

### MISCELLANEOUS POINTS

*Conveyance of adults on school buses.* Although this is now legal, arrangements for adults to use spare room on school contract buses have not been made widely. This is another example of failure to make best use of available resources. Allowing spare seats to be taken would help a substantial number of people, particularly in the most thinly-populated areas without a daily stage-service bus. Operators are sometimes reluctant to encourage application to the education authority lest term-time travel should lead to a demand for a bus at the same time during school holidays. It is sometimes argued that as school contracts are awarded annually, arrangements could only be made on a piecemeal, temporary basis. The Traffic Commissioners could most usefully bring the public, education authority and operator together in this matter.

*Mini-buses.* It is often suggested that operators could cut costs by using mini-buses on remote routes. In practice this is rarely a satisfactory solution. Traffic peaks are sharp, and where traffic is enough to justify a service at all, there are usually occasions on which more than a mini-busful of people want to travel. Small operators are naturally anxious to avoid destroying goodwill by leaving anyone behind. They say that mini-buses have to be regarded as extra vehicles rather than as replacements of 30-seaters, and that generally it is cheaper to use a 30-seater for both peak and slack periods than to alternate between two vehicles. Only when there is contract work for a mini-bus during the morning and evening peaks is its use on stage services during slack periods likely to be justified.

*One-man buses.* Wages of drivers and conductors account for about half the cost of bus operation, so using a joint driver and conductor on country routes is a worthwhile saving. After initial troubles, employees in most areas are now reconciled to the one-man system, though the substantial bonus which has to be paid to the driver for collecting fares obviously reduces the saving. Bus companies could have been more active in explaining the necessity for one-man operation to staff and also public, which sometimes objects to the change: often the alternatives are a one-man bus or no bus at all.

*Post Office buses.* It is sometimes maintained that adapting mail services to carry passengers as well would improve the rural transport position. But on existing routes peak traffic is usually too great for dual-purpose vehicles, which would therefore have to be restricted to areas at present without public transport. Post Office buses would inevitably run at times more convenient for mail than for passengers, and in most of Britain the majority of people would find it easier to hire a taxi, obtain a lift in a friend's car or walk to the nearest main-road stop. Only where distances are large and alternative means of

## Miscellaneous Points

transport negligible do Post Office buses with a few seats have any potential in this country. The Highlands and Islands of Scotland probably offer the only scope.

*Conveyance of mail by stage-service bus.* There is a much stronger case for the Post Office wherever possible to use existing bus services to carry mail. Sometimes this could be done without disruption to mail deliveries, but in sparsely-inhabited country (again notably the Highlands and Islands) some delay in mail might be justified if the passenger-carrying service were thereby preserved. Again, the Traffic Commissioners should help by encouraging co-operation. British Railways and British Road Services might more readily grant agencies for the conveyance of parcels and goods to outlying districts. A local garage might profitably combine passenger, mail and parcels services to and from the nearest centre, instead of separate authorities each making journeys at heavy loss.

*Bus trailers.* Another popular suggestion is that buses on some routes should be equipped with trailers for heavy luggage. In most cases the expense and difficulty of operation would outweigh the convenience, but there might be a future for trailers behind railway buses operated on a limited-stop basis over roads parallel to abandoned branch lines. On some routes more space should be provided inside the bus for luggage, even if this means sacrificing a few seats. A bus marked in the timetable 'Heavy luggage not conveyed' is not an adequate feeder service to trains.

*Bus shelters.* The public's standards are rising, and frequently the lack of shelters discourages traffic and increases motorists' temptation to offer lifts to waiting passengers. Most bus companies contribute only a little to the cost of shelters, and spread this over several years. A more generous approach would often be justified.

*Timetables.* Much could be done to help people discover bus times. Although some companies display departure times at important stops, others do not bother even in holiday areas. Printed timetables are sometimes difficult to obtain, not being available on the buses or at agencies. They rarely include the services of smaller operators in the same territory. In self-contained areas such as the Lake District, a comprehensive rail

## *Bus Services*

and bus timetable is badly needed. In the next few years the standard departure, or 'time interval', system—a bus at ten minutes past every hour, for example—may cease on many routes: traffic is declining far more rapidly at some times of day than others. The public will have to become more dependent on consulting timetables.

CHAPTER SIX

# TRANSPORT AND RURAL LIFE

### THE COST OF AVOIDING SUBSIDY

THERE are three main reasons why a broad network of rural transport services should be maintained, if necessary partly at the cost of the nation.

1. To be consistent with the policy of providing other rural amenities roughly up to town standards; to ensure that the capital spent on such things as sewerage, electricity, telephones and village halls is not wasted by the drifting away of the population for whom they were intended.

2. To help preserve a healthy balance of population between rural and urban areas, so that the countryside continues to play an active part in British life, and does not become a series of distressed areas and empty showpieces.

3. To enable a full life to be led by those still residing in the country without their own transport, even if they are the last generation to do so.

The three points of course overlap, but each has important aspects not shared by the others.

*1. Why exclude transport?* In Chapter 1, I spoke of the inconsistency of subsidizing most rural amenities without providing comparable help for transport. Here I can do no better than quote the report of the Council for Wales and Monmouthshire.

'The social and economic importance of the rural areas and the need to safeguard the welfare and living standards of people there have long been accepted as basic principles of social policy in this country. In accordance with these principles, such public services as water supply, sewerage, electricity, post and telephones, none of which can be economically provided in the rural areas, considered in isolation, have been made widely available throughout those areas with the aid of some form of internal or external financial help. . . . It has been, and still is, the aim of the responsible authorities to extend and maintain these services as widely as possible in the rural areas. Nor is there any question of withdrawing them, once they have been provided, on the grounds that they have become too uprofitable to continue. . . .

'There appears to be an inconsistency of attitude as between transport and other public services in the rural areas: in the case of the latter, it is recognized that the rural population should not be denied the everyday amenities of modern life notwithstanding that they cannot be economically provided; the provision and maintenance of public transport services, on the other hand, has come to depend more and more on considerations of accountancy, and travelling facilities which were at one time available to people in need of them are now rapidly disappearing. While it must be admitted that the analogy is not a precise one, largely because in the case of public transport alternative means of travelling are more readily available, it can be fairly argued that, for a significant proportion of people in the rural areas, public transport is still an amenity no less important than, say, a supply of electricity.

'Just as the absence of piped water and electricity supplies in the countryside have been matters of serious public concern, so, today, must the growing lack of railway and omnibus services in the rural areas give cause for deep anxiety. While it may be difficult to prove that the deficiencies in public transport services in the Welsh rural areas have had a direct influence on the stability of the population, there is a strong presumption that they have tended to reinforce the factors which cause depopulation. The withdrawal of local bus and train services cannot fail to have a depressing impact on the outlook of people in rural areas. The older generation inevitably feel a growing sense of isolation; the younger generation will be inclined to question, more seriously than before, whether there is a satisfying future for them in the rural areas. The economic development of the rural areas cannot but be hindered by deficiencies in the communications system and, in particular, measures taken to encourage the introduction of new industries, upon which rural authorities pin a great deal of faith as a means of halting depopulation, may be thwarted.'

An example of transport being the odd service out can be seen in mid-Wales. While the Government was stressing that no effort would be spared to stabilize the area's economy and to attract new sources of livelihood, the railways were sentenced

## The Cost of Avoiding Subsidy

to death for being uneconomic, and no thorough inquiry was undertaken to see how cheaply they could in fact have been run.

In many parts of rural Britain, an appreciable number of houses and cottages which have been modernized since the war, or have at least received electricity and piped water, now stand empty because people have moved to the towns. Inadequate transport has often been an important underlying cause, if not the obvious one.

The Council for Wales and Monmouthshire went on to point out that the continued drift of population from the Welsh rural areas was a matter of concern 'not limited to its effects on those areas. . . . The social disadvantages of the accumulation of population in the main industrial areas have long been publicly recognized.'

This, however, is an argument which apparently makes little impact when the future of transport services is officially considered. The need for rural electrification is admitted because this affects all country dwellers; public transport, on the other hand, is increasingly regarded as a luxury too expensive to provide for the decreasing number of people without cars and motor-bicycles. The attitude is that in future only those with a car will be able to live in the country. Apart from the human problem of established country people too old or too poor to run cars or to move, this overlooks the fact that public transport still remains important to many motoring families. People who always go by car themselves may rely on buses and trains to take their wives to the shops in mid-week and their children to work or school every day, once a week or only two or three times a year; and to bring the domestic help once or twice a week, and an ailing aunt for her annual visit. We travel how we please, but we expect public services to be available when required—including when we are ill, and in bad weather—and most of us are probably prepared to contribute to the cost, whether we use them much or not. Of course trains and buses cannot be provided *ad lib* for infinitesimal traffic, but is it any more rational to apply strict economic criteria here than to other rural amenities provided partly at the nation's cost?

2. *Does rural depopulation matter?* I strongly believe that it does matter. The question is well beyond the scope of the present

book, but without pretending to offer the full case, I should say that it has vital implications for the whole nation.

At present the rural economy is largely bolstered by the flimsy and outdated strategic argument that we need to produce an appreciable proportion of our own food. Yet over-production is likely to become increasingly embarrassing, and from the strict economic viewpoint probably the number of separate farms could be reduced by half, and output by a quarter or a third. The contraction would, however, have most serious psychological and other effects. The heart would fade from much of country life; market towns without substantial industrial or similar activities would be depressed almost as much as the villages; and rural counties would cease to exercise their valuable stabilizing influence on the national life. The countryside would of course become less attractive for touring and holidays. The value of having people living and *working* by the Scottish lochside, on the Yorkshire fells and in the Devon valleys cannot be calculated in money terms. I italicize the *working*, for a rural community loses character when what Professor Dudley Stamp calls its 'adventitious' population exceeds a certain proportion —when it becomes essentially a dormitory for the nearest town, or depends largely on tourism and retired people from other parts of the country instead of on agriculture, forestry, quarrying and other basic industries.[1]

But for the rapid swell in adventitious population on the urban or coastal fringes of a large number of rural districts, depopulation in the last few years would have appeared more startling. The detailed parish results of the 1961 Census are likely to produce a salutary shock, though even the Census hides the full story, for in parts of central and north Devon, for example, the present depopulation rate appears to be three or four times the average for the decade 1951–61. In passing, this is one reason why elsewhere I have suggested five-yearly censuses.

---

[1] The engaging question of 'what proportion of adventitious population can the countryside stand before it becomes a dormitory?' is discussed by R. J. S. Hookway in the *Journal of the Town Planning Institute*, July–August 1958. He quotes Clark's lecture on 'Planning Problems of the Countryside' given at the Town and Country Planning Summer School, 1948.

*The Cost of Avoiding Subsidy*

The adventitious population of course enjoys the amenities provided in the countryside at less than economic cost, but is generally not only the least necessary group for a healthy rural economy, but also the least dependent upon public transport. A network of public transport services is required for the benefit of the primary, indigenous population (especially wives and daughters).

*3. The Last Generation.* Supposing that the Government were officially to abandon the policy of shoring-up the rural economy and that more rapid depopulation were to become the official line, there would still be the human problem of making life as good as possible for those who remained. Many are too old to move without severe psychological hardship; many would lack the means to do so, especially since rural property values would fall. These, in the words of the Jack Committee, are 'those who cannot escape':

> 'In order to give some idea of the order of hardship involved, we have heard of housewives living in villages with no bus services within two or three miles and able to reach the local shopping centre only by hiring a taxi at considerable expense; of a woman with small children who had to push her pram some two miles along steep and narrow roads to the nearest bus; of young girls who on leaving school were unable to obtain employment because of the absence of a daily service to the nearest town, yet were too young to move into lodgings, which their starting wages would in any case not cover; and of elderly people able to leave their village only on rare occasions when they can afford to share a taxi. These are the cases where "hardship" seems a more appropriate description than "inconvenience". There are others which involve grave inconvenience; for example the shopping bus which allows either too long or not nearly enough in town. But we have formed the impression that the number of persons who are likely to experience this hardship may not be increasing, partly through the growth of private transport and partly through the migration of people to places where hardship will not be felt. On the other hand the hardship for those who cannot escape or who cannot have access to private transport may tend to increase. If any special provision is to be made, these are the places to which it should apply.'

Personally I think that some public transport in rural areas will remain important well into the twenty-first century, but taking the shortest viewpoint we cannot abandon the present generation. It is, indeed, largely the inconvenience and hardship endured by those who have no practical means of escape that prompted me to undertake this work. The way in which buses and trains are run cannot be regarded as merely the domestic concern of the operators when even a little reorganization of resources might enable country people to lead happier and fuller lives. When the problem is studied in its detail, the slowness of the Government to admit its seriousness, let alone take remedial action, seems inexplicable.

CONCLUSIONS REACHED FROM RESEARCH WORK

The conclusions reached from the research undertaken by the Lake District Transport Inquiry were summarized as follows:

1. The amount of open, positive hardship which can at the present time be attributed to inadequate transport is small.

2. There is, however, some hidden hardship. People do not generally complain of deprivations to which they have always been accustomed. Hidden hardship was particularly noticeable among the 15- to 20-year-olds, some of whom (in the households visited) were unable to take advantage of further education and other facilities open to town children. Probably only the use of motor-scooters could solve this.

3. Private-car ownership has resulted in increased total travelling but in a steady decrease in the use of buses. There were numerous examples of car owners giving their friends as well as relatives regular lifts.

4. People travel by private transport whenever possible, and even among those living in homes without a car there appears to be a growing 'consumers' opposition' to using buses, although this is difficult to measure. But to some extent the vast majority of the female population is still dependent upon public transport. A greater proportion of the journeys now made by bus are for essential purposes than would have been the case some years ago.

5. The large-scale withdrawal of bus services would cause hardship in many cases: even some of those who now travel by

## Conclusions Reached from Research Work

car for a large proportion of their journeys would feel seriously cut off were they unable to travel at all when their husbands had taken the car to work. Many men (commercial travellers and farmworkers, for example) would be pressed by their wives to leave their present rural homes.

6. Car ownership among women is spreading only slowly, and even in the younger age group a large proportion do not yet drive their husbands' cars. Dependence upon public transport *for at least some journeys* does not now appear to be declining rapidly. Not until the majority of women have their own personal transport will it be possible for bus services to cease in most villages without causing hardship and psychological if not actual isolation. The fear of having partially to *depend* on lifts in other people's cars is real.

7. In the meantime, though remaining fundamental to the rural economy's well-being, bus services will become steadily less used.

Broadly speaking these conclusions hold good for all the research work (in Devon, Northumberland and the Lake District) described in the second part of this book. The only significant modification, in fact, is that (point 1) less positive hardship was found already existing in the Lake District than in the other survey areas, notably the specimen village of Birtley in Mid-Northumberland.

Comparisons between Birtley and the other areas are drawn in Chapter 8, and to some extent may indicate what will happen elsewhere as the rural transport problem intensifies. Although the other specimen areas had a greater proportion of car owners than had Birtley, their population density was higher and the number of people dependent on public transport was also larger, so better bus services were still available. When services are reduced to the Birtley level, the amount of open, positive hardship among those still without private transport will rise—the point of course made by the Jack Committee and quoted on page 69.

I should therefore like to emphasize point 7. In the foreseeable future, bus services will remain necessary both to preserve a well-balanced village economy (with a strong primary population) and to prevent growing individual hardship. But the services will inevitably become more uneconomic, especially since a relatively small increase in car ownership may lead to a substantial drop in the number of passengers. This, in a nutshell, is the case for subsidy. The outlook for operators is gloomy, yet the village bus remains essential. Indeed, even the bus stop—or the stopping point where no actual shelter or post is provided—is an important institution. Operators justifiably claim that passengers are frequently given lifts in private cars because their presence at the bus stop indicated that they wish to travel—that on market days the same motorists may habitually pick up the same passengers at the same stops. Yet if the stops disappeared, the people concerned might prefer to stay at home rather than blatantly solicit a lift or set out with no assurance of obtaining one. Lifts are accepted eagerly where offered, but the fear of having to rely on them cannot be overstressed.

Universal car ownership cannot be the solution—certainly not during the period which is the concern of the Ministry of Transport and of town and country planning authorities. This was clearly demonstrated by the research work outlined in Chapter 8. True, it was found that two-thirds of the men aged between 35 and 54 owned a motor vehicle, but taking all age groups together there is still a long way to go before car ownership is universal even among able-bodied wage-earning men. Relatively few women yet drive their husband's car, in many cases because it is not available at the times required. The separate car for the wife is still almost entirely restricted to the upper income group and to women pursuing their own professional career.

'It is a fallacy to suppose that every rural dweller can be equipped with his own means of transport,' says the Council for Wales and Monmouthshire. 'In every village rural settlement there will inevitably be a proportion of people who cannot get about without the aid of public transport, either because they are old, young or disabled, or because they cannot afford, or for some other reasons, cannot make use of, private transport.'

A serious aspect of the problem is the inability of many young people to take advantage of further education and other training facilities, but the suggestion that scooters should sometimes be provided is frowned upon by most county councillors on social and safety as well as financial grounds. As bus services shrink, more school leavers will find themselves unable to train for the careers of their choice, without leaving home or taking their parents with them to somewhere more accessible. Already families are moving so as to be nearer their children's place of education or work—even families where father owns a car, but is not available to take his son (or more usually daughter) to the town.

### PLANNING AND TRANSPORT

At several points I have suggested what transport operators, sometimes in conjunction with strengthened Traffic Commissioners, might do to provide services more in keeping with demand. Just as important, town and country planning authorities should take a more active interest in public transport. There should be a lively two-way exchange: operators should be encouraged to provide the best possible service where development, or a halt to depopulation, is especially desirable, while the planning authority should take into account the well-being of transport services when considering development and other proposals.

Not even the most elementary consultations have taken place in most rural counties. New sewerage and building schemes are sanctioned with little if any reference to transport. Bus and train services are withdrawn with no reference whatever to the spirit or the letter of the area's development plan, which may be dislocated as the result.

Development plans should cover all aspects of a county's life and economy, but not even a passing allusion to ferries may be found in the plans of most of the counties where they still have their part in the transport picture. Bus services seldom figure in the list of village amenities. The railway system is usually described, but forecasts have often shown little understanding of the subject. As the result of forestry development, 'both passenger and goods traffic on the railway up the North Tyne valley will expand, though probably not beyond the capacity of the present single line', optimistically alleged the

Northumberland County Development Plan in 1952. Soon the line was completely closed.

The Forestry Commission have done much to bring new life and hope to rural districts, but their work would often have been yet more valuable had new settlements been built with regard to transport and other requirements. An excellent example of this occurs in the North Tyne area. Instead of adding to the village of Wark, on the main valley road, well served by buses generally under-patronized, the Commission built thirty-five houses on the inhospitable hillside at the end of the forest road at Stonehaugh, six miles away. Although (in 1959) only five people owned cars and three had motor-bicycles, there was insufficient support for a satisfactory new bus service. Typically, adults were not permitted to occupy empty seats on the school bus—and the Commission find it hard to fill their houses. (See pages 110–11.)

In areas such as central Devon where population and commercial and other resources are inadequate actively to maintain all present settlements, there is a strong case for allowing most of them to run down and concentrate fresh development at a 'key' village. Even a few oases of contemporary development could do much to enrich a declining countryside. But key villages can only succeed with the co-ordination of education, highway, transport and other services. Transport's role is vital.

The Minister of Housing and Local Government might usefully direct planning authorities to give fuller consideration to public passenger transport when next revising their development plans. The authorities should be invited to put their point of view on bus and train service changes or withdrawals, though formal protests are of slight value unless there are firm long-term proposals to be upset for the area concerned, or if the effect of closure is purely conjectural. Planning authorities should produce the evidence. It might be added that the closure of a branch line is the largest single change in land use which may be made without planning consent; logically the Ministry of Housing and Local Government should be involved in consultations before the Minister of Transport makes his decision.

CHAPTER SEVEN

# SUBSIDY AND OTHER ADMINISTRATION

## RECOMMENDATIONS IN REPORTS 1960-62

It might be useful to start with a summary of recommendations made in previous reports. They are in chronological order, beginning with my own *Rural Transport*.

*Rural Transport: A Report (1960).* Part or all of the money collected in fuel tax from stage-service bus operation should be remitted to the industry. A proportion, at least, of this sum should be set aside to help operators maintain rural services. The total money available for this purpose should be divided by the Minister of Transport between a number of regional committees, which would pay operators of selected routes on a mileage basis. The payments would depend on the committees' assessment of the circumstances. These committees would virtually be strengthened Traffic Commissioners, their own increased costs also being met from the fund formed by part of the remitted fuel tax. Each regional 'professional' committee would have attached to them a 'voluntary' Consumers' Council, interested in all forms of public passenger transport in their area and absorbing the present Transport Users' Consultative Committees.

*Rural Bus Services: Report of the Committee (1961).* (The Jack Committee Report.) All members supported the conclusion 'that adequate rural bus services cannot be provided except as a result of some measure of financial assistance from outside the industry'. The majority report dismissed a straightforward repeal of the fuel tax for all stage services as too sweeping and too imprecise a remedy. It also had the disadvantage of providing only temporary, 'once for all' help. Selective remission of tax

## Subsidy and Other Administration

to rural operators was dismissed as involving too many practical difficulties. Subsidy, the Report proposed, should be provided for new services and to maintain existing services which would otherwise be abandoned. Once it was decided to provide help for a service, this would be advertised with an invitation to operators to tender for it. The system would be administered by county councils, subsidy falling on the rates but attracting an Exchequer grant. The total cost of subsidy was estimated at £1,000,000 in the first year, 'with the prospect of some annual increase over the next five years'. (A quotation from the Report describing the proposed tendering system is included in the note on page 56.)

Three members produced minority reports, all disagreeing with the proposed tendering system. Mr. W. T. James said this could not work with the existing system of cross-subsidization. Especially if new operators obtained subsidy, many existing operators who now supplied unremunerative mileage might feel compelled to surrender their rural licences, which would end the valuable principle of cross-subsidization and would cost far more in subsidies than the majority Report concluded. Mr. James preferred a selective remission of fuel tax—a refund on stage-carriage mileage required to be operated under the terms of the operator's licences outside built-up areas. This would be administered by the Traffic Commissioners. 'Relief for rural services on these lines would depend wholly on matters of fact, and not upon the ebb and flow of local politics.'

Mr. H. R. Nicholas advocated the immediate abolition of the fuel tax on all stage-service mileage: 'there seems to me to be no case for granting a subsidy while at the same time extracting from the industry a heavy tax in respect of fuel consumed.' In the event of a total abolition of tax not being possible, abolition should be on a selected, rural basis (similar to that proposed by Mr. James). Mr. Nicholas also favoured the powers of the Traffic Commissioners being extended to administer the system.

Mr. Edward B. Powley took a somewhat similar line, believing that 'assistance can discriminatively and readily be given through the remission in full of the tax on fuel used on rural stage services only', and that this should be administered by the

## Recommendations in Reports 1960–62

Traffic Commissioners, not by county councils, who he thought would be opposed to accepting such responsibility.

*The Highland Transport Inquiry (1961).* The Inquiry disagreed with the Jack Committee's contention that subsidy should be administered by county councils and recommended that in the Highlands a single body or authority should be 'charged with the duties of determining whether a bus service is essential and the extent to which it requires financial assistance and of awarding contracts to enable essential services to be maintained.'

*The Council for Wales and Monmouthshire (1962).* Direct subsidy should be paid by the central Exchequer under a system administered by the Traffic Commissioners, not local authorities.

> 'Where no bus service existed along a particular route or notice had been given that an existing service was to be withdrawn, it would be open to any body or persons or local authority affected or any bus operator to apply to the Traffic Commissioners for a determination that a service was needed.'

After making full inquiries the Traffic Commissioners would announce their decision: there would be a right of appeal to the Minister. In the event of subsidy being agreed desirable, details of the service would be published and operators invited to tender for it. The Council also suggested that another committee might initially examine applications for a subsidized service: this committee would be set up generally 'to advise on positive steps which might be taken to achieve the better co-ordination and more effective use of existing transport resources in the rural areas.'

It is perhaps surprising that the principal report—that of the Jack Committee appointed to make recommendations to the Government on a national scale—should be that most at variance with the rest. It is the only one to suggest that organization of subsidy should fall to local government. The other reports, and the three minority reports of the Jack Committee, agreed unanimously that local transport would be better kept away from politics, a view generally taken also in public and

## Subsidy and Other Administration

Press discussion of the various reports. Another argument against using county councils, of course, is that many routes cross county boundaries.

It is, however, not merely the recommendation that county councils should determine the routes to be subsidized out of the rates, but that subsidy should be accompanied by a competitive tendering system, which has made the majority Jack Report so unpopular in the bus industry and unacceptable to most other experts. The almost universal practice of local authorities offering any form of contract is to accept the lowest tender, which would mean that under the Jack system subsidy would undoubtedly tend to be paid to small men, including new operators who might have sprung into existence specifically to claim the subsidy.

Mr. Powley complained that if the recommendations of the majority Jack Report were put into force, an established operator would be able to obtain subsidy only 'if willing to face surrender of his licence, gamble upon the likely action of the Rural District Council and County Council as to the need for the route, and risk unfair competitive quotation.' Mr. Nicholas specifically stated that the Traffic Commissioners should be given powers to grant subsidy 'using appropriate established operators as the media through which such a provision should be ensured'. The Highland Committee added a rider to their recommendations for subsidy: 'In awarding contracts regard should be had not only to the tenders offered but also to the development of the main arterial routes and the interests of the tourist industry since it may sometimes be that this can be done more effectively by the bigger company.' As expressed in *Rural Transport: A Report* and in Chapter 5 of the present work, my own view is that small operators could be encouraged to play a more active part than at present, but that the adoption of a straight tendering system would be unfair to established large operators and would result in rapid and piecemeal disintegration of the present rural bus network.

The Jack Committee took considerable pains to present a fair analysis of the problem in the countryside, but unfortunately touched only lightly on two fundamental issues following the conclusion that subsidy should be granted. Firstly, although

## A Professional and a Voluntary Committee

advocating arrangements which would lead to the expansion of small operators at the expense of the territorial bus companies, they did not attempt to show why the change might be beneficial. Indeed, the pros and cons of small operators were virtually ignored, and the Committee might have been unconscious of the implications of their recommendation. Secondly, the possibility of using the Traffic Commissioners to administer subsidy was rejected curtly. The sole comment is: 'Although this suggestion has certain attractive features we think it would be difficult to reconcile these additional responsibilities with those judicial functions which the Commissioners are appointed to discharge.'

### A PROFESSIONAL AND A VOLUNTARY COMMITTEE

The inability or reluctance of councillors to take a broad view instead of automatically accepting the lowest tender; the danger that applications would be considered against a political rather than an independent practical background; inconsistency between councils and difficulties over services crossing county boundaries; the fact that the counties whose transport services most urgently need aid would tend to be those most unwilling to add to the rate burden; these seem to me to be overwhelming arguments against requiring county councils to administer subsidy. Many county clerks, other officials, and councillors, recognize their force and would not welcome the extra responsibility.

The Traffic Commissioners, for all their lack of initiative in taking positive steps to improve the situation in present circumstances, have unique knowledge of the problems and potentialities. They are respected by public and operators for their independence of judgment. They alone have the experience to enable subsidy to be granted fairly without a strict competitive tendering system, and they alone are in a position to decide which individual services should be subsidized without incurring suspicion of political or other bias. The Jack Committee thought it would be hard to reconcile the administration of subsidy with the Commissioners' judicial functions, but could it not become precisely another judicial function?

## Subsidy and Other Administration

In the three years since writing *Rural Transport: A Report*, I have been constantly in touch with the rural transport problem and can still suggest no better solution for bus services than that outlined in the second paragraph of this chapter, although experience prompts amendments of detail and amplification of some of the proposals.

In the first place, then, money to help rural bus services should come from the fuel tax. The origin of the money, however, is relatively unimportant. What matters is the creation of a fund to help rural transport, and if a single fund could be employed to subsidize bus services and individual branch railways, and to pay the higher administration costs of the Traffic Commissioners, so much the better. The size of the national fund and its distribution between road and rail services, and between the different regions, would be political decisions and the direct responsibility of the Minister of Transport, who might however be advised (in the same way as he is now by the Central Transport Users' Consultative Committee) by a central committee including representatives of the two kinds of regional committees I propose should be established. The day-to-day administration of bus subsidy within each region would be decided locally, though operators and the public would have the right of appeal to the Minister. Subsidy for individual branch railways thought worth retaining in the national interest might be the responsibility of the Minister advised by the new Traffic Commissioners.

*The New Traffic Commissioners.* The first of the regional committees would virtually be strengthened Traffic Commissioners, and for convenience may be referred to as the new Commissioners. They would usually cover slightly smaller areas than do the present Commissioners. They would employ more staff, while a number of their members, appointed by the Minister, should be industrialists and others with administrative experience paid on a part-time basis.

As already said, the new Commissioners would be allocated by the Minister an annual sum to spend on bus subsidies in their area, and arguments about the size of this sum would be directed at the Minister and the Government. Within their areas the new Commissioners should be able to distribute the

## A Professional and a Voluntary Committee

fund according to the merits of each route as established by their own investigations without becoming the centre of serious political controversy.

Generally, I believe, subsidy for bus services should be granted on a mileage basis. With research facilities at their disposal in addition to their accumulated experience, the Commissioners could establish a fair rate of payment. Cross-subsidization would not automatically cease; operators serving rich territory would still have to accept some rural losses without help. The new Commissioners should also be strong enough to encourage and even enforce better use of small operators without threat to the well-being of the territorial bus companies. Competitive tendering for subsidized services could be introduced if felt desirable in particular cases, but the way in which the licensing system has been worked by the Traffic Commissioners during the last thirty-two years suggests that rarely would this course be taken.

The new Traffic Commissioners would issue licences as at present; they would consider complaints and suggestions coming from the Consumers' Committee (to be mentioned in a moment); with their own research officer, they would occasionally initiate proposals on matters besides subsidy, and in particular would be charged with obtaining fuller co-operation between road, rail and other forms of passenger public transport. They would take an active interest in railway matters, though primarily in an advisory capacity. When British Railways wished to close a branch line, for example, the Commissioners would report to the Minister on the way it was being run and the extent to which its costs might be cut.

I firmly believe that such new Commissioners would work to the benefit of the transport industry as well as of the public, and that they would win the confidence of both operators and passengers. Although they would be far more active than are some of the present Commissioners, and would have considerably augmented powers, the day-to-day running of buses would of course remain entirely in the hands of the operators.

*The Consumers' Committee.* The new Traffic Commissioners which I am proposing would in each region have attached to them a purely advisory committee, consisting of voluntary

## Subsidy and Other Administration

members, mainly from local government. Their chairman would sit with the new Commissioners, while the Commissioners' secretary might also act for them. This Consumers' Committee would be concerned with all forms of public transport. They would consider complaints and suggestions from the public and initiate suggestions for improvements. They would be entitled to discuss the application of subsidy to bus services in their area and to make recommendations to the Commissioners.

The Consumers' Committee would absorb the responsibilities of the present Transport Users' Consultative machinery, except that either the Commissioners and the Consumers' Committee would hold a joint inquiry into objections to a proposal to close a branch line, or they would make their separate recommendations to the Minister. In the latter event the Consumers' Committee would confine themselves to recommendations on the score of hardship, the Commissioners being responsible for reporting on matters of cost and operation. In Chapter 4 I stressed that the present Transport Users' Consultative Committees were weakened by their close railway affiliations. The Consumers' Committee would be entirely independent, financed through the Minister, not directly by the transport industry.

### THE COST OF SUBSIDY

The rural transport problem could be substantially solved, at least for 10–20 years, by the expenditure of a sum less than that paid in fuel tax by all stage-service buses. I would recommend the allocation of about £5,000,000 for the purpose—perhaps an initial £3,000,000 rising to a ceiling of £5,000,000 in five years.[1]

Not only is £5,000,000 a modest enough sum in itself, but a considerable part of it would be justified by direct savings to the education authorities, Post Office and other organizations. Moreover, as already stated, some outside financial support for rural transport is necessary to ensure that other subsidized rural amenities are not wasted. Suggestions that subsidy

[1] The Jack Committee were told that in 1958 the Tilling group of companies alone spent £3,130,000 on fuel tax for their stage services, excluding express and contract buses.

## The Cost of Subsidy

granted for specific purposes would prove extremely costly for nebulous results are unwarranted. Even the strictly economic arguments in favour of it are strong, quite apart from the question of human happiness.

Of the £5,000,000, between £1,000,000 and £2,000,000 would be spent on subsidizing bus services (excluding those in the Scottish Highlands), chiefly on a mileage basis. The Jack Committee felt that initially £1,000,000 would be sufficient. But the system they put forward would have led to the rapid drying-up of cross-subsidization, and in those circumstances Mr. James is almost certainly correct in his forecast that the cost would be 'probably many times greater' than estimated. £1,000,000 might, however, well prove sufficient if arrangements of the kind I have suggested were implemented, because the bus companies would continue to run many routes at a loss without subsidy, and where subsidy were granted it would not necessarily completely bridge the gulf between cost and revenue. Thus generally no subsidy might be paid on routes making an average loss of less than 6d. a mile, and 6d. a mile might be paid if the loss were 9d.—and it would be the duty of the new Commissioners to check that the bus company had taken reasonable measures to achieve the best results. Spread out at 6d. a mile, for example, £1,000,000 a year would cover five daily journeys each way six days a week on 1,000 routes each thirteen miles long.

Where an adequate case was presented, subsidy would be available to start new, or restart abandoned, bus routes, but of committees and individuals investigating the problem in recent years, the Jack Committee were alone in believing that there was a considerable demand for additional routes. Some sort of bus service has generally survived so far where the demand is anything like sufficient to justify one. The urgent task is to maintain services at their present severely-reduced level; the cost of serving areas not now on the transport map would usually be excessively high in relation to the social benefit bestowed.

A sum not exceeding £3,000,000 would be used to subsidize or otherwise improve the profitability of individual branch lines (excluding those in the Scottish Highlands) worth retaining in

## Subsidy and Other Administration

the national interest. Generally subsidy would be granted for a given number of years to span, or partially to span, the estimated gap between income and expenditure, allowance being made for the value of through traffic brought to the main line by the branch trains. Again, the new Commissioners would ensure that reasonable steps had been taken towards making the services economic before recommending the Minister to grant a subsidy.

In some cases, however, an outright grant might be paid to enable the railways to undertake capital works—such as the adaptation of signalling—to reduce daily running costs. The expenditure of a few hundred pounds could sometimes save substantial operating costs, as discussed in Chapter 4, and if a branch is to be retained purely for social reasons the necessary capital should not be expected to come from the Railway Board.

Subsidy would also be available where appropriate to support railway-controlled buses replacing branch or stopping trains and running specifically in connection with main-line services.

Initially about £1,000,000 might be spent on annual subsidies to keep lines open, £1,500,000 on grants for capital works, and a few thousands to help establish the first railway-controlled bus services. The amount needed for grants would tend to decline, and subsidies to rise, after a few years. As a rough guide, according to British Railways' own estimates the annual loss on many single-line branches could be cut to £500–£1,000 a mile if the simplest possible methods of operation were employed and the service reduced to the basic minimum. In these circumstances £1,000,000 would cover about 1,300 route miles.

These estimates do not include provision for services for the Scottish Highlands and Islands, where of course some mail-boat services are already partially maintained by direct grant. For various reasons we may concur with the Highland Transport Inquiry that the Highlands and Islands are a special case, though the basic machinery just suggested should work there as elsewhere in Britain.

Finally, allowance has to be made for the cost of the new

## The Cost of Subsidy

Commissioners and of the Consumers' Committee I have proposed. In each region, salaries and the remuneration of part-time members of the Commissioners would not exceed £12,000 and office and other costs £6,000. Supposing that Britain were divided into twelve regions, each with Commissioners and an attached Consumers' Committee, the total cost would be about £200,000, from which has to be deducted the present cost of operating the Traffic Commissioners and the Transport Users' Consultative Committees.

# PART TWO

CHAPTER EIGHT

# RESEARCH WORK

## THE VILLAGE SURVEYS

THE most important task was the detailed interviewing of a cross-section of country people about their transport habits. In the first year of research, a questionnaire was put to virtually every resident over the age of 15 in the parishes of Ashton and adjoining Trusham in the Teign Valley in Mid-Devon (261 interviews) and in the parish of Birtley on the North Tyne in Mid-Northumberland (253 interviews). In these cases we completed forms for every person on the Electors' List who was still living in the villages, and also for young people over 15 in the houses we visited.[1]

As part of the Lake District Transport Inquiry, similar questions were put to a cross-section of residents in ten parishes: both the parishes and the people were chosen by recognized random-sampling methods. This was a rural project and towns and large main-road villages were deliberately avoided. We set out to interview sixty people in each of the parishes and completed 533 interviews. The discrepancy was due to deaths and moves, and occasionally to other inaccuracies in the Electors' Lists, but a few people were away from home throughout the period of the Inquiry and in four cases we could not obtain reliable information. The Lake District figures are comparable with those from Ashton–Trusham and Birtley, except that in the Lake District, where we were not interviewing the whole population in any one place, we had to confine ourselves to people on the Electors' Lists and did not therefore include anyone in the 16–20 age group.

In total the travel habits of 1,047 people are reflected in the

[1] At Birtley we slightly stretched the parish boundary so as to give a total sample of the same size as at Ashton and Trusham combined.

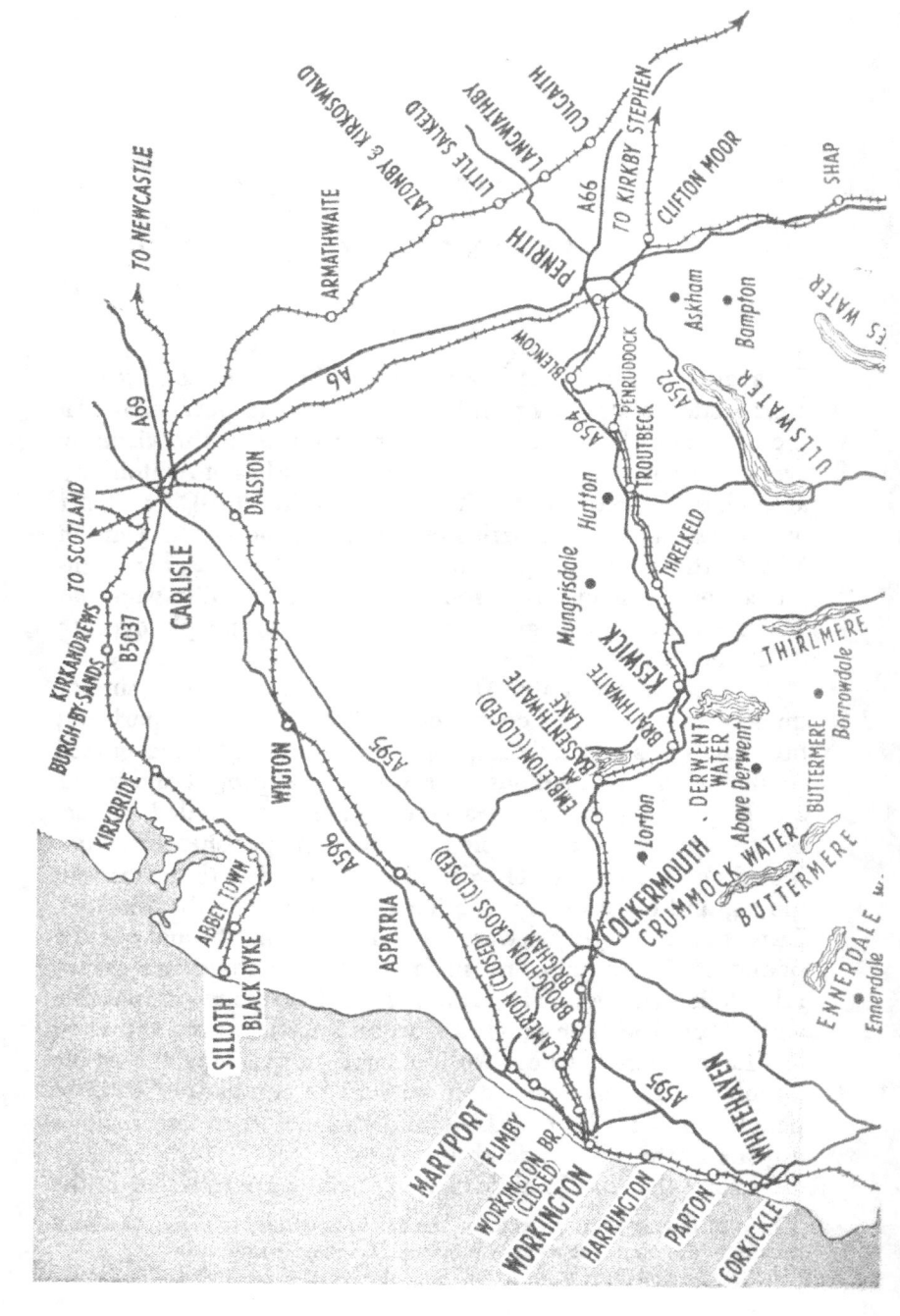

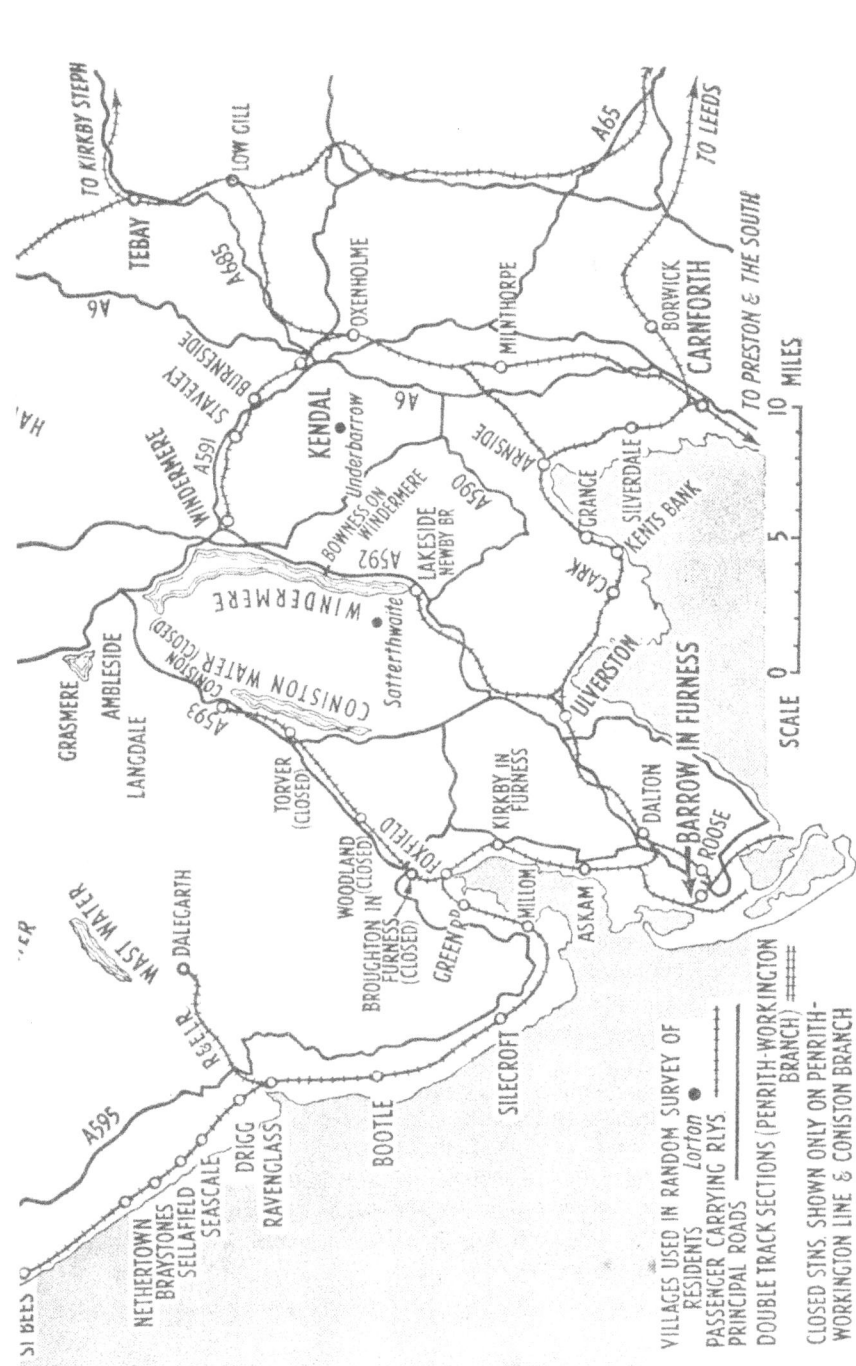

The Lake District showing the survey villages of the Lake District Transport Inquiry

# THE LAKE DISTRICT SURVEY VILLAGES
*(as found at time of survey)*

*Key to Letters:* C, Cumberland; L, Lancashire; W, Westmorland; c, a fair amount of commuting to nearest towns; f, forestry fairly important; p, population (1931 and 1951 respectively); r, a fair number of retired people from other areas; t, tourism fairly important. The letters in italics refer to common complaints (not those made by only a few people), the most frequent complaint being shown first. The letters denote difficulty in reaching: *d*, doctor; *e*, cinema and dances; *l*, long-distance trains; *s*, shops.

ABOVE DERWENT, C, r, t. Only part of parish was covered, excluding main-road villages. The rural part is served by bus three times daily, passing within half a mile of most homes. *e, d.*

ASKHAM, W, p 444, 386, c, t. A compact village mainly dependent upon Lord Lonsdale's estates. Bus service about four times daily to Penrith is operated by Mr. E. Hartness and loses heavily. *d, e, l.*

BAMPTON, W, p 1,011, 407, c, r, t. Population decline largely due to Manchester's reservoir which drowned part of parish. Water works now give some employment. Served by same bus service as Askham. *e, l.*

BORROWDALE, C, p 581, 724, c, r, t. The main road is served by a joint bus service frequent in summer, run by Cumberland Motor Services and a private operator. Considerable building activity. *l, e.*

ENNERDALE AND KINNISIDE, C, p 331, 281, f. Apart from the small village of Ennerdale Bridge, the population is scattered. *d, s.*

HUTTON, C, p 368, 306, c, f. A scattered parish mainly of small owner-occupied farms. Penruddock and Troutbeck stations are within the parish, and main-road bus service is good. Few complaints made here.

LORTON, C, p 272, 265, c, f, r, t. Most of the population is at High Lorton, a compact village. *l, e, s (in winter).*

MUNGRISDALE, C, p 392, 371. A hillside village two miles north of Keswick–Penrith road. One-third of population is in village centre, where only three households are without a car. A market-day bus to Penrith has been withdrawn. *d, s, e, l.*

SATTERTHWAITE, L, f. In the little-visited area between Lakes Windermere and Coniston. There is a twice-weekly bus. *d, l, s, e.*

UNDERBARROW, W, p (1951) 314. A scattered parish without village served by bus three days a week. *d, l, e.*

## The Village Surveys

'village surveys'. We posted questionnaires in advance, and received adequate replies to about one-third. The rest of the interviews were conducted personally. Except for the under-20s in Ashton–Trusham and Birtley, all the interviews were with people on pre-determined lists: as the very subject was transport, we could not drop somebody who was repeatedly out. There was no substitution.

As might be expected, the last 15 per cent of the lists for each parish accounted for over half the time and cost: to complete one interview, we had to open gates across lanes over thirty times. The work took us into some of the most beautiful country, but also along some of the worst lanes and drives, in England. It brought us intimately in touch with the rural population, and was as valuable for its background information as for the facts and figures we were actually seeking.

It is not claimed, of course, that the sample is representative of the whole of rural Britain, although it does include all types of people living in all types of villages. The proportion of retired people who have moved from towns is probably higher than in the countryside generally. Table 3 shows the employments of the population of each of the three areas. In all areas forestry, quarrying and transport were the chief local sources of employment besides farming.

The centre of Birtley is two miles from Wark, on the North Tyne. Wark was on the Border Counties Railway, from Hexham to Riccarton Junction, closed in 1956, and is served by a good daily bus service on the Hexham–Bellingham road. Birtley itself is served only by a twice-weekly bus to Hexham.[1] It was formerly a retainers' village, and each cottage in the single short street has its small allocation of land. There is a church, school, inn, shop and separate post office, but each of these is seriously under-employed: most of the parish's population live outside the village, and the majority of business is transacted at the small market centre of Bellingham or at Hexham, the capital of Mid-Northumberland. In the outlying parts of the parish are several large estates with cottages for domestic staff, several medium-sized farms employing shepherds, and numerous small

[1] Although the present tense is used, unless otherwise stated facts are those found at the time of the survey.

farms relying entirely on family labour. Hardly any houses have been built this century, and many of the more isolated cottages have fallen into disrepair or been taken over as week-end homes by Tyneside people. Almost all the permanent population is of local origin.

Ashton and Trusham were two stations on the Teign Valley Railway from Exeter to Heathfield, which had connections to Newton Abbot, and closed in 1958. British Railways have subsidized the Devon General Omnibus Company to run a bus service along the main Teign Valley road, but this is inferior to the former train service. Ashton is a two-part village, the lower half, just off the main road, being the more important: there is a shop, public house and smithy. Higher Ashton, a mile out of the valley towards Haldon Moor, includes the parish church. Most of the indigenous population are farmers or farm or forestry workers and their families, but a number of professional and retired people have moved in during recent years and some of them (or their children) commute to Exeter or Newton Abbot. Trusham is larger and more compact and has a wider range of facilities. But it is over a mile up a narrow road from the Teign Valley and has never had its own bus service. A number of the men work at the quarry and concrete works in the valley, and there is considerable commuting to Newton Abbot.

The Lake District survey was intended to be representative of the rural Lake District as a whole. Compared with ninety-four farmers in the survey there are only sixteen farm workers: most Lakeland farms are small family affairs. There are as many employees in forestry as in agriculture. The proportion of retired people is high. Among the 'other employed' are a number of salesmen and commercial travellers, some of whom take their cars out of the area throughout the working week, a sprinkling of office and hospital workers (mainly young women), and several in the catering industry. A summary of details about the ten parishes chosen will be found on page 92.

*Vehicle Ownership: Tables 1–3*

Of a total of 195 men aged between 35 and 54 in the three surveys combined, no fewer than 124 own vehicles. This in itself seems almost to sum up the problem facing transport

## The Village Surveys

operators in rural areas. To quote the individual figures (Table 2), in Ashton-Trusham 28 out of 40 own vehicles, in Birtley 30 out of 59, and in the Lake District 66 out of 96.

Men aged between 35 and 54 might well be regarded as the core of the population. If two out of three of them own cars, this proportion may be expected to extend steadily to a larger age range. The rising standard of living will help younger men to buy a vehicle earlier than their fathers could, while many who have run a car in middle age will continue to do so well into retirement. Far fewer cars were owned, of course, when the present old men were middle-aged: indeed in country districts there is still a substantial remnant of the last generation who regard motor vehicles as a novelty.

The tables show a marked contrast between Ashton-Trusham and the Lake District on the one hand and Birtley on the other. In the first two cases over a third of the adults own vehicles; in Ashton-Trusham, in fact, less than a third of those interviewed live in households which include no vehicle owner. In Birtley, however, only about one person in five owns a vehicle, and half the population live in households without any motor transport.

This might be expected, Birtley being chosen to represent the poorer type of strictly rural parish. But it is significant that it is among the men aged 35–54 that the proportion of car owners comes nearest to that in the other two surveys. Because of poorer public transport, a car is perhaps more urgently needed here than in the other survey areas, and many of these men will buy one at considerable sacrifice. But few of the retired have money to spare. Among the young men, vehicle ownership is restricted by two important factors: Birtley's climate is unsuitable for motor cycles, and some young men working on their fathers' farms do not receive a full wage. In the whole of Birtley only three women own a car—three out of 129, compared with 9 out of 135 in Ashton-Trusham and 38 out of 286 in the Lake District.

Although the present trend of car ownership among the middle-aged probably indicates the future pattern among the young and the old, it is of course no reliable guide. Many factors have to be taken into account. For example, it seemed that the average age of cars was higher in Birtley than elsewhere, and fewer might last into their owners' retirement.

## TABLE ONE
### General Summary

|  | Ashton | Birtley | Lake District | Total |
|---|---|---|---|---|
| Own cars | 70 | 47 | 169 | 286 |
| Own motor-cycles | 27* | 7 | 14* | 48 |
| No personal motor transport but in car-owning household | 81 | 65 | 159 | 305 |
| No motor transport owned personally and no car in household | 83 | 134 | 191 | 408 |
| Total | 261 | 253 | 533 | 1,047 |

* In addition five car-owners in Ashton and two in the Lakes also own motor-cycles.

## TABLE TWO
### Ownership of motor vehicles

| Age groups | Under 21 M | F | 21–34 M | F | 35–54 M | F | 55 and over M | F | Total |
|---|---|---|---|---|---|---|---|---|---|
| *Total of three samples* | | | | | | | | | |
| Total | 23 | 17 | 101 | 110 | 195 | 204 | 178 | 219 | 1,047 |
| Car owners | 2 | — | 56 | 11 | 112 | 22 | 67 | 16 | 286 |
| Motorcycle owners* | 6 | 5 | 16 | 3 | 12 | — | 6 | — | 48 |
| *Ashton* | | | | | | | | | |
| Total | 13 | 13 | 33 | 26 | 40 | 44 | 40 | 52 | 261 |
| Car owners | 1 | — | 19 | 1 | 24 | 6 | 17 | 2 | 70 |
| Motorcycle owners* | 6 | 5 | 9 | 2 | 4 | — | 1 | — | 27 |
| *Birtley* | | | | | | | | | |
| Total | 10 | 4 | 15 | 17 | 59 | 64 | 40 | 44 | 253 |
| Car owners | 1 | — | 8 | — | 26 | 2 | 9 | 1 | 47 |
| Motorcycle owners | — | — | — | — | 4 | — | 3 | — | 7 |
| *Lake District* | | | | | | | | | |
| Total | — | — | 53 | 67 | 96 | 96 | 98 | 123 | 533 |
| Car owners | — | — | 29 | 10 | 62 | 14 | 41 | 13 | 169 |
| Motorcycle owners* | — | — | 7 | 1 | 4 | — | 2 | — | 14 |

* In addition five car-owners in Ashton and two in the Lakes also own motor-cycles.

## The Village Surveys

### TABLE THREE

*Occupation of vehicle owners*

|  | 1 Own cars | 2 Own motor-cycles* | 3 Car in house | 4 Own pedal-cycle | 5 No trans-port | Total |
|---|---|---|---|---|---|---|
| *Total of three samples* | | | | | | |
| Farmers | 109 | 1 | 23 | 4 | 28 | 165 |
| Farmworkers | 13 | 12 | 13 | 7 | 11 | 56 |
| Forestry | 10 | 6 | — | 9 | 6 | 31 |
| Mining, quarrying | 14 | 5 | 2 | 1 | 9 | 31 |
| Transport | 17 | 3 | — | 9 | 9 | 38 |
| Other employed | 65 | 20 | 57 | 22 | 38 | 202 |
| Housewives, retired, unoccupied | 58 | 1 | 210 | 26 | 229 | 524 |
| Total | 286 | 48 | 305 | 78 | 330 | 1,047 |
| *Ashton* | | | | | | |
| Farmers | 23 | — | 3 | — | 4 | 30 |
| Farmworkers | 6 | 7 | 4 | 2 | 2 | 21 |
| Forestry | 3 | 2 | — | 5 | 2 | 12 |
| Mining, quarrying | 8 | 3 | 1 | — | 3 | 15 |
| Transport | 7 | 2 | — | 2 | — | 11 |
| Other employed | 13 | 13 | 14 | 7 | 3 | 50 |
| Housewives, retired, unoccupied | 10 | — | 59 | 4 | 49 | 122 |
| Total | 70 | 27 | 81 | 20 | 63 | 261 |

\* Those who own both cars and motor-cycles are classified as car-owners.

(1) Owners of cars.

(2) Owners of motor-cycles.

(3) No personal motor transport but in car-owning household.

(4) Live in household without car and own personally only a pedal-cycle.

(5) No motor transport owned personally and no car in household.

## TABLE THREE—continued

*Occupation of vehicle owners*

| | 1<br>Own cars | 2<br>Own motor-cycles* | 3<br>Car in house | 4<br>Own pedal-cycle | 5<br>No trans-port | Total |
|---|---|---|---|---|---|---|
| *Birtley* | | | | | | |
| Farmers | 24 | — | 7 | 1 | 9 | 41 |
| Farmworkers | 3 | 2 | 2 | 4 | 8 | 19 |
| Forestry | 1 | — | — | 1 | 1 | 3 |
| Mining, quarrying | 4 | 2 | — | 1 | 3 | 10 |
| Transport | 2 | — | — | 7 | 3 | 12 |
| Other employed | 9 | 3 | 12 | 10 | 14 | 48 |
| Housewives, retired, unoccupied | 4 | — | 44 | 6 | 66 | 120 |
| Total | 47 | 7 | 65 | 30 | 104 | 253 |
| *Lake District* | | | | | | |
| Farmers | 62 | 1 | 13 | 3 | 15 | 94 |
| Farmworkers | 4 | 3 | 7 | 1 | 1 | 16 |
| Forestry | 6 | 4 | — | 3 | 3 | 16 |
| Mining, quarrying | 2 | — | 1 | — | 3 | 6 |
| Transport | 8 | 1 | — | — | 6 | 15 |
| Other employed | 43 | 4 | 31 | 5 | 21 | 104 |
| Housewives, retired, unoccupied | 44 | 1 | 107 | 16 | 114 | 282 |
| Total | 169 | 14 | 159 | 28 | 163 | 533 |

\* Those who own both cars and motor-cycles are classified as car-owners.

(1) Owners of cars.
(2) Owners of motor-cycles.
(3) No personal motor transport but in car-owning household.
(4) Live in household without car and own personally only a pedal-cycle.
(5) No motor transport owned personally and no car in household.

## The Village Surveys

But the Lake District figures would not appear so different from Birtley's were it possible to disregard the retired people and the professional and other workers who have moved in from other areas, attracted by the Lakes setting. Indeed, that Birtley is not untypical of the poorer rural Northcountry was borne out by very similar results from the two upland parishes surveyed in the Lake District. Incidentally, in the Lake District and in Birtley we found that about one-third of the farmers were without cars. Some of these farmers have car-owning relatives living with them, and most of their acreages are small.

### Travel Frequencies: Tables 4–7

Relatively few of those in our combined sample in three areas make daily journeys but four-fifths travel at least once a fortnight to at least one destination. As might be expected, vehicle owners travel more frequently and to a greater variety of places than do the rest of the population.

Table 4 shows that the use of public transport for daily journeys is virtually confined to the 'other employed' in which there is a high proportion of young people, especially girls. These include the daughters of farmers and farm and forestry workers, who have secretarial, shop and other jobs in the nearest town. Many of the 'other employed' who travel daily by private transport are also young: in a number of cases several use the same car, or even two the same motor cycle. The daily journeys of farmers and of farm and forestry workers are usually short, to work or about the estates, although a few take children to and from school. The table does not include journeys on pedal cycles.

Interesting differences between Birtley and the other areas emerge again. For example, Table 5 shows that the number of daily travellers at Birtley is little more than a third that at Ashton–Trusham, yet nearly three times as many travel between one and four times a week. This, it seems, is because the weekly shopping expedition remains an extremely important social event in Northumberland, and bus patronage on market days increases more dramatically than elsewhere. Homes and gardens generally supply more interest for Devon people, and many without cars go shopping only when really necessary, not for enjoyment.

One-fifth of the total of the three samples (but an insignificant number of car owners) travel less often than once a fortnight to any one destination. The 157 who travel less often than once a fortnight, and the 52 who never travel, include many old people and invalids, but especially in the Lake District we found an appreciable number of able-bodied women who seldom if ever visit the nearest town because of transport difficulties, and also a few men who seemed never to travel—mainly from choice. Not all cars have heavy use: in fact in two cases they were kept purely for journeys within a radius of a mile or so, and the owner of another holds no driving licence and merely meets tradesmen and children at the end of his long drive.

Table 6 shows that three-quarters of the total sample regularly visit one or two places, while about one-fifth divide their travel between three or more destinations and the remainder do not go to any one place regularly.[1] Motorists not only travel more often but divide their regular journeys between more destinations than do those relying on public transport. But among motorists, we found that those who make the most journeys in total are not usually those who regularly visit the greatest number of different places. There was some difficulty in defining 'place': the difference between the top line of this table and the bottom line of Table 5 is due to the fact that a small number of people who make regular journeys go to destinations which we could not easily classify. In one case, for example, a motorist who normally uses his car only on and around his own land has to visit a petrol station several miles away about once in ten days.

About 60 per cent in the combined three surveys make some use of public transport, and 89 out of 334 vehicle owners do so, although on average less frequently than other people. It seems that people in households without cars use public transport

---

[1] These figures may be untypical of the rural population as a whole, for both Birtley and Ashton–Trusham have market centres on either side and a high proportion of their residents go to both.

## The Village Surveys

about five times as often as car owners.[1] Comparisons may be made between Table 7, concerned only with journeys by public transport, and Table 5, which shows all journeys. The question concerned regular even if occasional use, and the answers are analysed by the destination most often visited: infrequent long-distance journeys by public transport to a variety of holiday destinations are therefore not included, although a regular annual trip to the same resort may be.

TABLE FOUR

*Daily journeys*

|  | Ashton | | Birtley | | Lake District | |
|---|---|---|---|---|---|---|
|  | Public transport | Private | Public transport | Private | Public transport | Private |
| Farmers | — | 3 | — | 2 | 1 | 12 |
| Farmworkers | — | 2 | — | 2 | — | 3 |
| Forestry | — | 6 | — | 2 | — | 6 |
| Mining, quarrying | — | 9 | 1 | 3 | — | 6 |
| Transport | — | 8 | — | 2 | 2 | 5 |
| Other employed | 10 | 18 | 3 | 9 | 11 | 38 |
| Housewives, retired, unoccupied | — | 4 | 1 | 1 | 4 | 10 |
| Total | 10 | 50 | 5 | 21 | 18 | 80 |

[1] This may be worked out by estimating the actual frequency of the journeys, multiplying this frequency by the number of people concerned, and taking into account the total number in each category. The estimate can be made rather more accurately as a comparison between those in car and non-car households by using the sample of bus travellers described in note on page 119.

## TABLE FIVE
*Frequency of travel, by destination most visited*

|  | Car and motor-cycle owners | No personal vehicle but in car-owning household | No personal vehicle and no car in household | Total |
|---|---|---|---|---|
| *Total of three samples* | | | | |
| 1. 5 or more days a week | 104 | 26 | 44 | 174 |
| 2. 1–4 days a week | 106 | 74 | 95 | 275 |
| 3. Week to fortnight | 103 | 139 | 147 | 389 |
| 4. Less often | 18 | 46 | 93 | 157 |
| 5. Never | 3 | 20 | 29 | 52 |
| Total | 334 | 305 | 408 | 1,047 |
| *Ashton* | | | | |
| 1. 5 or more days a week | 38 | 10 | 11 | 59 |
| 2. 1–4 days a week | 15 | 13 | 10 | 38 |
| 3. Week to fortnight | 35 | 42 | 34 | 111 |
| 4. Less often | 9 | 14 | 24 | 47 |
| 5. Never | — | 2 | 4 | 6 |
| Total | 97 | 81 | 83 | 261 |
| *Birtley* | | | | |
| 1. 5 or more days a week | 11 | 3 | 9 | 23 |
| 2. 1–4 days a week | 27 | 34 | 38 | 99 |
| 3. Week to fortnight | 14 | 17 | 40 | 71 |
| 4. Less often | 1 | 7 | 42 | 50 |
| 5. Never | 1 | 4 | 5 | 10 |
| Total | 54 | 65 | 135 | 253 |
| *Lake District* | | | | |
| 1. 5 or more days a week | 55 | 13 | 24 | 92 |
| 2. 1–4 days a week | 64 | 27 | 47 | 138 |
| 3. Week to fortnight | 54 | 80 | 73 | 207 |
| 4. Less often | 8 | 25 | 27 | 60 |
| 5. Never | 2 | 14 | 20 | 36 |
| Total | 183 | 159 | 191 | 533 |

(1) People who travel to one or more places five or more times a week.
(2) People who travel to one or more places at least once a week.
(3) People who travel somewhere between once a week and once a fortnight.
(4) People who travel less frequently than once a fortnight.
(5) People who never travel.

## TABLE SIX
*Places regularly visited*

|  | Car and motor-cycle owners | No personal vehicles but in car-owning household | No personal vehicle and no car in household | Total |
|---|---|---|---|---|
| *Total of three samples* | | | | |
| No places | 7 | 24 | 38 | 69 |
| One place | 113 | 94 | 149 | 356 |
| Two places | 126 | 119 | 156 | 401 |
| Three and over | 88 | 68 | 65 | 221 |
| Total | 334 | 305 | 408 | 1,047 |
| *Ashton* | | | | |
| No places | 4 | 4 | 5 | 13 |
| One place | 31 | 16 | 26 | 73 |
| Two places | 32 | 43 | 37 | 112 |
| Three and over | 30 | 18 | 15 | 63 |
| Total | 97 | 81 | 83 | 261 |
| *Birtley* | | | | |
| No places | 1 | 4 | 14 | 19 |
| One place | 7 | 12 | 39 | 58 |
| Two places | 18 | 23 | 56 | 97 |
| Three and over | 28 | 26 | 25 | 79 |
| Total | 54 | 65 | 134 | 253 |
| *Lake District* | | | | |
| No places | 2 | 16 | 19 | 37 |
| One place | 75 | 66 | 84 | 225 |
| Two places | 76 | 53 | 63 | 192 |
| Three and over | 30 | 24 | 25 | 79 |
| Total | 183 | 159 | 191 | 533 |

## Research Work

### TABLE SEVEN
*Use of public transport*

| Destination most visited | Car and motor-cycle owners | Car in house | | No car in house | Total |
|---|---|---|---|---|---|
| | | drivers | non-drivers | | |
| **Total of three samples** | | | | | |
| 5 or more days a week | 4 | 1 | 9 | 18 | 32 |
| 1-4 days a week | 5 | 3 | 27 | 80 | 115 |
| Week to fortnight | 24 | 10 | 71 | 142 | 247 |
| Less often | 56 | 16 | 53 | 97 | 222 |
| Never | 245 | 34 | 81 | 71 | 431 |
| Total | 334 | 64 | 241 | 408 | 1,047 |
| **Ashton** | | | | | |
| 5 or more days a week | — | — | 6 | 4 | 10 |
| 1-4 days | 1 | 1 | 2 | 5 | 9 |
| Week to fortnight | 9 | 2 | 25 | 32 | 68 |
| Less often | 19 | 3 | 16 | 21 | 59 |
| Never | 68 | 10 | 16 | 21 | 115 |
| Total | 97 | 16 | 65 | 83 | 261 |
| **Birtley** | | | | | |
| 5 or more days a week | — | — | 1 | 3 | 4 |
| 1-4 days | — | 1 | 14 | 24 | 39 |
| Week to fortnight | 3 | 2 | 8 | 42 | 55 |
| Less often | 19 | 5 | 13 | 46 | 83 |
| Never | 32 | 3 | 18 | 19 | 72 |
| Total | 54 | 11 | 54 | 134 | 253 |
| **Lake District** | | | | | |
| 5 or more days a week | 4 | 1 | 2 | 11 | 18 |
| 1-4 days a week | 4 | 1 | 11 | 51 | 67 |
| Week to fortnight | 12 | 6 | 38 | 68 | 124 |
| Less often | 18 | 8 | 24 | 30 | 80 |
| Never | 145 | 21 | 47 | 31 | 244 |
| Total | 183 | 37 | 122 | 191 | 533 |

## The Village Surveys

*Motor Cycles and Pedal-Cycles*

Motor-cycles and scooters provide an important part of the transport for the not-so-wealthy in Ashton–Trusham. At Ashton we found four girls aged between 20 and 22 commuting by scooter. Weather conditions in the Lake District and even more in Mid-Northumberland limit the use of motor-cycles, and we found that five of the seven motor-cycle owners in Birtley regularly visit Hexham by bus, and only three do so by motor-cycle. Of these three, only one never uses the bus to Hexham. Even in Ashton–Trusham, motor-cycle owners use public transport relatively often—much more often than do car owners. Many motor-cycles are owned mainly for getting to work and are not much used for week-end social outings with friends or relatives.

We were told of 43 pedal-cycles in Birtley, 46 in Ashton–Trusham and 42 in the Lake District, but many are used rarely if at all. Those in good condition are mainly kept for short journeys—including daily journeys to work—and are regarded as alternatives to walking rather than to motor vehicles. Again, the weather severely limits use of bicycles in the North, especially in Birtley. We found that women who owned bicycles used public transport more often than the average.

*Transport Difficulties: Table 8*

At the end of the interview, the subjects were asked if they had any particular transport difficulties, and the tables analyse the results from the three areas. The figures reflect the difference between the areas, but in every case the largest number of people complained of the difficulty of reaching the doctor's surgery. Shopping is not included in the list because few people had serious trouble in obtaining the necessaries of life, even though they complained of the lack of choice at the shops they could reach. A total of ninety-two people in the three surveys complained of the difficulty of reaching main-line trains, either to travel themselves or to meet friends.

Those we interviewed were also asked if they had any special comment to make. At Birtley the largest number of remarks—twenty-four—concerned the cost of hiring taxis. Eighteen people mentioned that they relied on lifts in other people's cars. There were numerous suggestions for improved bus services, especially on Sundays, and two people said that conditions would be much easier if they had permission to travel on a school bus which had spare seats. The prices charged by travelling shops and their poor selection of goods were the complaint of a number of people in the remoter areas, and several said that the shops could not reach them in winter when the lanes were muddy. These were among the other comments: 'I would move to Bellingham if I could', 'I'm lonely in Birtley so have taken a job elsewhere', and 'The journey to Hexham takes so long that now I spend three nights a week there'. Only one employer complained that inadequate transport was causing difficulty in attracting workers—much of the surrounding area is even more distant from bus services—but a number grumbled that the early evening buses which brought workers home from Hexham were too late for women shoppers needed to prepare their husbands' teas.

In Ashton–Trusham most of the remarks concerned the buses. Twenty-five said that at least an occasional bus should go up to Trusham village and to Higher Ashton, and twenty-five complained of poor connections between the Newton Abbot–Leigh Cross and Exeter–Christow services at Leigh Cross, and about the lack of a shelter. Twenty complained of the lack of late evening services on Saturdays, and two stated that they had bought a vehicle chiefly for Saturday night expeditions, although having bought it they used it for all journeys. In all, twelve stated that they had recently bought motor vehicles because of the worsening transport position. Fifteen complained of the expense of hiring taxis when they wanted to begin a long-distance journey by train, or meet friends at a station, and seven said that they relied on lifts for such journeys. Five families stated that they were moving as soon as possible because of transport troubles: two because the husband could no longer reach his work without difficulty, two because they felt generally cut off, and one in order to give their daughter better chances of further education. Other remarks included: 'Without a car I should be immobile', 'A car is too expensive but you can't do without it here', 'Car

## The Village Surveys

or own transport considered necessary in this village', 'I am teaching others scootery', and 'You feel a prisoner here'. It should perhaps be mentioned again that Ashton-Trusham lost their railway service in 1958, and many of the complaints referred to the fact that the bus service has not adequately replaced the trains. More complained of the difficulty of travelling with prams here than elsewhere: in the Lake District, by contrast, people have never been able to transport large prams and did not complain of the fact.

### TABLE EIGHT

*Transport difficulties*

| Age group: | 15–20 | 21–34 | 35–54 | 55 and over | Total |
|---|---|---|---|---|---|
| *Ashton* | | | | | |
| Football | — | 2 | 2 | 2 | 6 |
| Cinema and dances | 5 | 5 | 9 | 6 | 25 |
| Friends | 3 | 5 | 12 | 12 | 32 |
| Main-line trains | — | 4 | 8 | 23 | 35 |
| Prams | — | 7 | 3 | 1 | 11 |
| Doctor's surgery | 3 | 13 | 16 | 24 | 56 |
| Other reasons | 2 | 2 | 4 | 5 | 13 |
| *Birtley* | | | | | |
| Football | — | — | 5 | 3 | 8 |
| Cinema and dances | 3 | 2 | 8 | 6 | 19 |
| Friends | 2 | 4 | 15 | 9 | 30 |
| Main-line trains | — | 4 | 11 | 10 | 25 |
| Prams | — | — | 1 | — | 1 |
| Doctor's surgery | 2 | 3 | 33 | 23 | 61 |
| Other reasons | — | 2 | 6 | 3 | 11 |
| *Lake District* | | | | | |
| Football | — | 3 | 1 | 3 | 7 |
| Cinema and dances | — | 12 | 10 | 14 | 36 |
| Friends | — | 10 | 8 | 11 | 29 |
| Main line trains | — | 5 | 14 | 13 | 32 |
| Prams | — | 4 | — | 1 | 5 |
| Doctor's surgery | — | 9 | 18 | 22 | 49 |
| Other reasons | — | 10 | 10 | 20 | 40 |

*Research Work*

*Women's Driving Habits: Tables 9 and 10*

Few country wives have their own car, and the number does not seem to be rising rapidly. The use women make of their husbands' cars is therefore a matter of importance.

In the village surveys we asked all women not owning a car themselves but living in a household in which somebody else has one whether they drove it, and if not we asked them the reason. We supplemented this information by two other means. A separate questionnaire was put to all women not themselves on the village survey list but living in car-owning homes visited in the course of the village survey in the Lake District. And the questions about driving were included in the questionnaires filled in at meetings of the Westmorland Women's Institutes (see next section).

Table 10 shows that in the basic village surveys only 39 drive compared with 216 who do not drive the household car. The figure of 216 includes some who do not—naturally enough—drive the car of a lodger or someone else living at their home other than the husband, but even allowing for this only a small proportion of wives drive their husbands' cars. As might be expected the highest proportion of drivers are in the Teign Valley, followed by the Lake District, with Birtley again the least 'advanced' area.

Table 11 combines the reasons given in all five inquiries for not driving the household car. Several gave two answers and some preferred to give none. Probably the 'husband dislikes' reason was soft-pedalled, especially when husbands were present at the interviews. Both the 'husband dislikes' and the 'too nervous' answers reflect a social attitude which is probably breaking down less quickly in rural than in urban areas. Even in the survey areas these reasons are less important among the younger women and more of the younger women do actually drive. Though slow, change is taking place.

It should be stressed that even if all wives without their own personal vehicle drove their husbands' cars, most would have to rely on public transport for at least part of the week. Sixty-four stated specifically that they had not learnt to drive because their husbands take the family car to work, and in many other

## The Village Surveys

### TABLE NINE

*Women who do not personally own cars; numbers who drive in car-owning households*

|  | Age group: | Under 21 | 21–34 | 35–54 | 55 and over | Total |
|---|---|---|---|---|---|---|
| Ashton | Drive | — | 7 | 5 | 1 | 13 |
|  | Don't | 8 | 10 | 19 | 22 | 59 |
| Birtley | Drive | — | 2 | 5 | — | 7 |
|  | Don't | 2 | 7 | 27 | 10 | 46 |
| Lake District |  |  |  |  |  |  |
| Random sample | Drive | — | 9 | 7 | 3 | 19 |
|  | Don't | — | 32 | 46 | 33 | 111 |
| Supplementary interviews | Drive | — | 16 | 11 | 4 | 31 |
|  | Don't | — | 24 | 42 | 35 | 101 |
| Westmorland Women's Institutes | Drive | 6 | 51 | 84 | 25 | 166 |
|  | Don't | 9 | 35 | 184 | 94 | 322 |

### TABLE TEN

*Total of five surveys: Reasons for not driving given by non-car owners living in car-owning households*

| Age group: | Under 21 | 21–34 | 35–54 | 55 and over | Total |
|---|---|---|---|---|---|
| Driving lessons too expensive | 4 | 7 | 21 | 2 | 34 |
| Too nervous | 3 | 26 | 131 | 52 | 212 |
| Husband dislikes | — | 8 | 50 | 36 | 94 |
| Father dislikes | 5 | 3 | — | — | 8 |
| Too old | — | 2 | 18 | 77 | 97 |
| Too young | 14 | — | — | — | 14 |
| Husband takes car to work | — | 12 | 44 | 8 | 64 |
| Now learning | — | 2 | 4 | 1 | 7 |
| Other reasons | 9 | 42 | 67 | 31 | 149 |
| Total* | 23 | 112 | 322 | 210 | 676 |

\* Of which 37 were men. A few gave no reason and some gave more than one reason.

## Research Work

cases this is no doubt a contributing factor. If the car would not be available when required, there is little point in women considering whether they are really afraid to drive.

*Supplementary Inquiries*

To obtain a broader picture of the support for public transport in the Lake District, we arranged with the Westmorland Federation of Women's Institutes for questionnaires to be filled in by members attending a regular monthly meeting of each branch in the county. Some 1,060 completed questionnaires were analysed—about as many as for the three village surveys together. The same basic pattern emerges: some car owners sometimes use public transport; some of those who do not own a car personally but live in a household in which a car is owned still rely heavily on buses, though they use them less than do women from homes without any motor transport. The sample is obviously not representative, for Women's Institutes attract a more than averagely mobile section of women to their meetings. The proportion of car owners is much higher than found among women in the Lake District village survey; it is likely that W.I. members use buses more often than most women, but there may possibly have been a tendency to exaggerate bus use 'to give a good showing' at meetings.

As an extension of the village surveys, we went to Stonehaugh, a Northumberland Forestry Commission village, which we knew was untypical. The results are interesting in themselves and are of the greatest importance so far as town and country planning is concerned, but they have not been added to those from the survey village proper. Stonehaugh is in a wild area six miles from Wark in the opposite direction to Birtley. There are 35 occupied houses, each with a man and wife, and there are 5 children aged over 15. Only 5 own cars, 3 have motor-cycles and 1 a pedal-cycle. Only 4 women work, 1 of these being the sole daily traveller out of the village, and only 16 leave the village to go to the same destination at least once a week. For financial reasons the vehicle owners use buses for half their journeys. Out of 75 people, 44 never use private transport, and the twice-weekly bus service is thinly supported.

Nearly all the men are on basic agricultural pay, and as

## The Closure of Two Branch Railways

already stated only 4 out of 35 wives are wage-earners. Many of the families have come from Tyneside, attracted by the offer of a modern house, but there is a rapid turnover of population: 25 of the 75 people have lived there less than two years, and only 29 more than five years. Even if money were not a problem, few Stonehaugh people would find much to attract them in Wark or even Hexham, but many pay an occasional visit to friends in Tyneside. Though the bus is little used, its withdrawal would make it even harder for the Forestry Commission to keep a contented labour force, but such transport help as the Commission gives has the effect of lessening the demand for the bus. There is also a school bus. This has empty seats, but adults are not allowed to use it.

### THE CLOSURE OF TWO BRANCH RAILWAYS

We decided to find out what had happened to the travel habits of the regular users of two closed railway lines in Devon—the Teign Valley and the Moretonhampstead branches, which joined each other at Heathfield.

Lists of regular passengers living in the villages served (not passengers travelling there from elsewhere) were collected on the trains before closure. They included everyone who made daily journeys, and a high proportion of those making journeys at least once a fortnight. It was not so easy to discover those who travelled only infrequently. Between nine and twelve months after the closures, people on the lists who were still living in the villages were interviewed personally.

*The Teign Valley Branch: Table 11*

The Teign Valley line between Exeter and Heathfield was closed in June 1958. The trains had faced competition from buses between Exeter and Christow, but provided the only public transport at Ashton and Trusham. Chudleigh is served by two bus routes, and its inconveniently-situated station was little used. When the trains were withdrawn, British Railways subsidized the Devon General Omnibus Company to run a new bus service from Newton Abbot to Leigh Cross, connecting with the Christow–Exeter service. There are fewer buses than

## TABLE ELEVEN

### Effects on former users of closure of Teign Valley railway

| | Use of train | | | | Journeys over same route one year after closure | | | | | Bought motor vehicle since closure | Travel less often by public transport elsewhere |
|---|---|---|---|---|---|---|---|---|---|---|---|
| | daily | weekly | less often | | as frequent | less frequent | not made | by bus | by other means | | |
| Heathfield | 1 | — | 2 | a<br>b | —<br>— | —<br>— | —<br>1 | —<br>— | —<br>— | —<br>— | —<br>— |
| Chudleigh | 3 | 1 | 1 | a<br>b | 2<br>2 | 1<br>— | 1<br>1 | 1<br>1 | —<br>1 | —<br>— | 2<br>1 |
| Trusham | 5 | 11 | 5 | a<br>b | 4<br>10 | —<br>5 | —<br>1 | 2<br>15 | —<br>2 | —<br>2 | 3<br>5 |
| Hennock and* Teign Village | 3 | 7 | 6 | a<br>b | 1<br>5 | —<br>7 | 2<br>1 | 1<br>10 | —<br>2 | —<br>— | 1<br>3 |
| Ashton | 9 | 10 | 2 | a<br>b | 9<br>2 | —<br>9 | —<br>1 | 5<br>10 | 4<br>1 | 4<br>1 | 5<br>2 |
| Canonteign† | 2 | 4 | 1 | a<br>b | 1<br>1 | 1<br>4 | —<br>— | 1<br>5 | —<br>1 | —<br>— | —<br>— |
| Christow | 7 | 5 | — | a<br>b | 6<br>4 | —<br>1 | —<br>— | 4<br>4 | 2<br>1 | 1<br>— | 3<br>4 |
| Doddiscombsleigh‡ | 2 | 1 | 2 | a<br>b | 4<br>2 | —<br>2 | —<br>— | —<br>2 | 2<br>1 | 2<br>1 | —<br>1 |
| Total | 32 | 39 | 19 | a<br>b<br>a & b | 25<br>25<br>50 | 1<br>29<br>30 | 6<br>4<br>10 | 14<br>48<br>62 | 12<br>6<br>18 | 9<br>2<br>11 | 15<br>15<br>30 |

*a* Daily users. *b* Others. \* Passengers used Trusham station. † Passengers used Ashton station.
‡ Passengers used Christow station.

## The Closure of Two Branch Railways

there were trains, and the buses do not run to the railway station at Newton Abbot.

Of 90 regular users of the Teign Valley branch, 32 used to travel every day and 39 at least once a week. The inquiry revealed that a year after closure only one-third of the sample were using the bus service as much as they had used the railway. Roughly one-third, including nearly half the daily travellers, had deserted public transport in the area altogether, either having bought a motor vehicle or through changed circumstances. (For example, several who had travelled by train to work, but who had always used private transport at week-ends, obtained local jobs after the closure.) The remaining third of the sample did use the buses, but less frequently than they had used the trains.

The table analyses separately what happened to the former daily travellers. Of the 32, nine immediately bought motor vehicles, 2 arranged lifts and 5 lost their jobs. We found that road journeys (by car or bus) nearly all cost more than the rail fares. But only 24 of the 90 regular passengers said the closure had caused them 'great inconvenience'; 47 complained of 'some inconvenience'; the rest said they had suffered 'no real inconvenience'. Only 15 thought that ending the service had been right, and many suggested that British Railways should have changed to diesel trains and advertised the route as a pleasant alternative to the main line between Exeter and Newton Abbot. Specific complaints were headed by the difficulty of reaching main-line trains (including those to Torbay): no fewer than 26 of the 90 people mentioned this: 20 mentioned the difficulty of meeting friends, 19 of shopping, 18 of reaching the doctor, 16 of going to the cinema, and 6 to dances. Allied to the last two difficulties, 8 complained of the lack of a late bus on Saturdays. The biggest number of complaints about the bus service (14) concerned the lack of shelters, especially at Leigh Cross.

*The Moretonhampstead Branch: Table 12*

The Newton Abbot–Heathfield–Moretonhampstead line was closed in March 1959. There was bus competition throughout the route except that the buses did not leave the main road to serve Lustleigh: after withdrawal of the trains, British Railways subsidized the Devon General to call there. The railway had the advantage for Lustleigh and for Moretonhampstead, reached by train from Newton Abbot in half the time taken by bus, and the largest number of passengers interviewed had used these two stations. Altogether we interviewed 166 people, of whom 55 travelled daily. Nine months after the closure, only 77 of the 166 were making their journeys as often by road (bus or car) as they had done by rail; 73 were making the journeys less frequently, and 16 not at all: 120 made some use of the bus service, but only 49 used it as often as they had the trains, and of these 29 did so daily to reach their work. Of the other 26 who used to go daily by train, 9 (mainly women, including domestic servants) were going by bus but had reduced their jobs so that they travelled on certain weekdays only, 6 had bought motor vehicles, 3 had lifts in other people's cars, 1 walked, and 7 had changed their jobs (6 specifically because of the railway's closure, and 1 for another reason). Including the non-daily users of the branch, 11 former travellers had bought cars. These figures all exclude schoolchildren who of course travelled as before.

Of the remaining 111 passengers (those who did not travel daily), 13 said they had ceased travelling altogether and 62 were doing so appreciably less often. Of those who travel as often as before, over a third do so by car. Many women said they visited Newton Abbot less frequently for shopping, doing more business in their villages or at the door: 13 mentioned mobile shops. No fewer than 72 said that they had visited the seaside less: in the first (fine) summer after closure, only an estimated quarter of the previous annual train trips to the sea were made by bus.

Only 7 people thought that British Railways had been right to close the branch: 43 said that the closure had caused 'great inconvenience'; 71 said it had caused 'some inconvenience' and the rest 'no real inconvenience'. But 88 said that as the result of the closure they used other public transport (outside the Moretonhampstead–Newton Abbot district) less often, 47 complained of the lack of connections between the buses and main-line trains at Newton Abbot, and 13 said they had to use taxis to reach the main railway. Twenty-two said that a much-reduced train service with tickets issued by the guard would

TABLE TWELVE

*Effects on former users of closure of Moretonhampstead branch*

| | Use of train | | | | Journeys over same route one year after closure | | | | | Bought motor vehicle since closure | Travel less often by public transport elsewhere |
|---|---|---|---|---|---|---|---|---|---|---|---|
| | daily | weekly | less often | | as frequent | less frequent | not made | by bus | by other means | | |
| Newton Abbot | — | 2 | 1 | a | — | — | 1 | 2 | — | — | — |
| | | | | b | — | 2 | — | — | — | — | — |
| Teigngrace | 1 | 2 | 2 | a | 1 | — | — | 1 | — | — | 2 |
| | | | | b | — | 2 | — | — | — | — | — |
| Heathfield | 6 | 3 | 2 | a | 4 | — | — | 4 | — | — | 4 |
| | | | | b | 4 | 2 | — | 6 | — | — | 1 |
| Brimley and Bovey Tracey | 6 | 5 | 9 | a | 2 | 3 | — | 5 | — | — | — |
| | | | | b | 6 | 7 | 2 | 6 | 1 | — | 4 |
| Pullabrook | 4 | 5 | 2 | a | 5 | 2 | — | — | 4 | — | 9 |
| | | | | b | 2 | 3 | 1 | 11 | 4 | — | 2 |
| Lustleigh | 5 | 23 | 10 | a | 3 | — | — | 2 | 1 | 1 | 5 |
| | | | | b | 5 | 15 | 2 | 4 | 7 | 1 | 3 |
| Moretonhampstead | 13 | 22 | 14 | a | 16 | 2 | 1 | 24 | 2 | 2 | 15 |
| | | | | b | 10 | 27 | 4 | 10 | 3 | 2 | 8 |
| North Bovey* | 7 | 3 | 4 | a | 5 | 3 | 1 | 29 | 3 | 1 | 26 |
| | | | | b | 3 | 4 | 2 | 3 | 1 | 2 | 4 |
| Chagford* | 13 | 2 | — | a | 1 | 2 | 1 | 4 | — | 1 | 2 |
| | | | | b | 10 | 1 | 1 | 8 | 4 | 1 | 3 |
| Total | 55 | 67 | 44 | a | 41 | 10 | 3 | 38 | 13 | 6 | 28 |
| | | | | b | 36 | 62 | 13 | 82 | 17 | 5 | 60 |
| | | | | a & b | 77 | 73 | 16 | 120 | 30 | 11 | 88 |

*a* Daily users.   *b* Others.   \* Passengers used Moretonhampstead station.

have sufficed, and another 12 said that even an occasional train would have solved their troubles, particularly those of reaching the main line. In addition to those mentioning difficulty in reaching the main line, 9 spoke of the difficulty of visitors reaching them. Other remarks were: route by Lustleigh and Heathfield too indirect (34, mainly Moretonhampstead people), bus more expensive than train (33), feel ill on buses (29, again mainly Moretonhampstead people—it seemed a fashionable thing to say), unable to travel with prams (13), no bus shelters (12).

We followed up the closure of three other lines in detail, although with less complete coverage. These were the Ashburton branch in Devon, the Border Counties line in our Mid-Northumberland survey area, and the Coniston branch in the Lake District. Exactly the same pattern was found as with the Teign Valley and Moretonhampstead lines. Only about one-third of the journeys made by residents of the villages concerned were transferred to the alternative bus service, and in all five cases there was an even greater drop in tourist and other traffic from towns to branch-line destinations. In total probably in none of the five cases was more than 20 per cent of the railway's passenger business transferred to the stage-bus service. This does not of course necessarily assert hardship: some who used to go to villages merely to see the countryside by train now do so by motor-coach excursion or by car, and many go to other destinations without feeling aggrieved. That was why we interviewed only residents of the villages when we set out to measure inconvenience and hardship.

A STUDY OF BUS PASSENGERS

As part of the Lake District Inquiry, passengers were interviewed on a typical selection of bus journeys. At least one journey on most routes in the survey area (excluding industrial hinterlands) was covered, and more journeys were covered on busy routes than on quiet ones.[1] The interviewers questioned

[1] It was of course impossible to be exactly representative: routes served only once or twice a week had to be over-represented to be included at all. Our staff were working awkward hours on other assignments and were not free to cover fully early morning and late evening buses, although the evening up to 7 p.m. was well represented.

## A Study of Bus Passengers

every passenger on the bus, but the whole journey was not always covered. In the summer study period, 1,008 passengers were interviewed, 349 of them on strictly country routes and the rest on routes linking towns such as Keswick and Penrith. In winter a further 402 were questioned.

Residents accounted for 60 per cent of the summer passengers and visitors 40 per cent. As might be expected, there were no visitors on some journeys, and on others they predominated, but summer visitors seem to make some use of nearly all routes. In winter—during the Christmas school holidays—visitors accounted for only 20 per cent of passengers and were mainly confined to two routes. Day trippers from outside the Lake District form an insignificant part of bus traffic.

### The Residents: Tables 13 and 14

The summer and winter questioning produced similar results, and the two samples may be considered together. The outstanding feature was the high proportion of women passengers —641 compared with 206 men.

To reach conclusions about the effects of car ownership on bus travel, we need to know what proportion of the population in the area have cars in their household. For this we may go back to the Lake District village survey; although some of the bus travellers interviewed were town people, a substantial majority came from areas similar to those of the village survey. The village survey concerned only people over the age of 21, and the youngest age group of bus passengers has therefore been omitted for this comparison:

|  | Men | Women | Total |
|---|---|---|---|
| Percentage of bus passengers over 21 years old with car in house | 21 | 32 | 29 |
| Percentage in village survey with car in house | 66 | 58 | 62 |

Clearly those with cars in their households use buses less frequently than those without, and this applies particularly to men. The estimated 38 per cent of the adult population without cars in their households (village survey figure) provide 71 per

## TABLE THIRTEEN

*Lake District bus users' survey: Residents' combined summer and winter figures; frequency of travel by bus*

|  | Under 21 | | 21–34 | | 35–54 | | 55 and over | | Total |
|---|---|---|---|---|---|---|---|---|---|
|  | Car† | No Car | Car† | No Car | Car† | No Car | Car† | No Car |  |
| **Men:** | | | | | | | | | |
| 5 or more days a week | 9 | 6 | 7 | 16 | 4 | 14 | — | 14 | 70 |
| 1–4 days a week | 3 | 4 | 3 | 7 | 3 | 18 | 3 | 37 | 78 |
| Less often | 3 | 4 | 10 | 7 | 6 | 12 | 1 | 15 | 58 |
| Total | 15 | 14 | 20 | 30 | 13 | 44 | 4 | 66 | 206 |
| **Women:** | | | | | | | | | |
| 5 or more days a week | 21 | 11 | 10 | 17 | 10 | 38 | 9 | 23 | 139 |
| 1–4 days a week | 7 | 10 | 30 | 39 | 68 | 71 | 25 | 117 | 367 |
| Less often | 2 | 3 | 14 | 12 | 11 | 38 | 9 | 46 | 135 |
| Total | 30 | 24 | 54 | 68 | 89 | 147 | 43 | 186 | 641 |
| **All:*** | | | | | | | | | |
| 5 or more days a week | 30 | 18 | 17 | 34 | 15 | 56 | 9 | 37 | 216 |
| 1–4 days a week | 11 | 15 | 35 | 46 | 71 | 93 | 28 | 159 | 458 |
| Less often | 5 | 7 | 25 | 20 | 17 | 53 | 10 | 64 | 201 |
| Total | 46 | 40 | 77 | 100 | 103 | 202 | 47 | 260 | 875 |

\* Includes some for which sex is not specified.
† A car in the household, irrespective of ownership.

cent of bus traffic, which means that on average people from homes without cars travel by bus about four times as frequently as do the rest of the population.

Table 13 shows, as might be expected, that an appreciable number of young people travel daily. People under 21 account for a quarter of all daily travellers, but make only a small contribution to less frequent travellers. In total we estimate that about 4 per cent of the adult population use buses

## A Study of Bus Passengers

### TABLE FOURTEEN

*Lake District bus users' survey: Residents' summer and winter figures; purposes of travel*

|  | Business‡ | Shopping | Entertainment | Friends | Medical | Other |
|---|---|---|---|---|---|---|
| Men: |  |  |  |  |  |  |
| Car in house* | 28 | 9 | 8 | 7 | 2 | 1 |
| No car in house | 77 | 42 | 19 | 28 | 2 | 2 |
| Women: |  |  |  |  |  |  |
| Car in house | 51 | 137 | 14 | 31 | 5 | 1 |
| No car in house | 100 | 237 | 29 | 74 | 7 | 2 |
| Total† | 269 | 445 | 72 | 140 | 16 | 7 |

\* A car in the household irrespective of ownership.

† Includes some for which sex is not specified.

‡ Includes a few children on their way to school, but buses on which children predominate were avoided. The figures also include several women who travel by bus to take their young children to and from school.

daily, with a further 20 per cent travelling at least once a week.[1]

[1] This may be checked by comparing the sample of bus passengers and the results of the village survey. Table 13 shows 216 passengers travelling five or more times a week, 458 on one to four days a week, and 201 less often. If we estimate the frequency in the first two groups as about five and a half times and just under twice a week respectively, we should expect to find six or seven times as many people in the second group as in the first. This is reasonably consistent with the village survey figures (given below) of those using public transport analysed by the destination most frequently visited. The village survey of course took train travel into account but in the village chosen almost all journeys by public transport would have been by bus.

| | |
|---|---|
| Travel five or more days a week | 18 |
| Travel one to four days a week | 67 |
| Travel once a week or fortnight | 124 |
| Travel less than once a fortnight | 80 |
| Never use public transport | 244 |
| | 533 |

## Research Work

Table 14 shows the principal purposes for which the bus passengers were travelling. Sometimes more than one reason was stated, joint shopping and entertainment trips being fairly common. Even allowing for the small number of evening buses covered it seems that entertainment trips now account for only about 10 per cent of total use. Shopping accounts for 47 per cent of bus use in our sample, and business for not quite a third (29 per cent). Many women who need buses for shopping trips, made during working hours, can travel in private cars at weekends and sometimes in the evenings, and it is probable that among residents bus use will become yet more strictly confined to shopping and business.

### The Visitors: Tables 15 and 16

Although the summer and winter questioning of residents pro-

**TABLE FIFTEEN**

*Lake District bus users' survey: Visitors' summer figures; frequency of local travel during holidays*

| Frequency | Under 21 | 21–34 | 35–54 | 55 and over | Total |
|---|---|---|---|---|---|
| **Men:** | | | | | |
| 5 or more days a week | 4 | 12 | 21 | 21 | 58 |
| 1–4 days a week | 9 | 18 | 20 | 7 | 54 |
| Less often | 5 | 4 | 5 | 2 | 16 |
| Total | 18 | 34 | 46 | 30 | 128 |
| **Women:** | | | | | |
| 5 or more days a week | 10 | 25 | 42 | 35 | 112 |
| 1–4 days a week | 9 | 23 | 25 | 23 | 80 |
| Less often | 5 | 9 | 3 | 4 | 12 |
| Total | 24 | 57 | 70 | 62 | 213 |
| **Total:*** | | | | | |
| 5 or more days a week | 19 | 39 | 66 | 59 | 183 |
| 1–4 days a week | 29 | 50 | 51 | 31 | 161 |
| Less often | 13 | 13 | 8 | 6 | 40 |
| Total | 61 | 102 | 125 | 96 | 384 |

* Includes some for which sex is not specified.

## A Study of Bus Passengers

### TABLE SIXTEEN

*Bus users' survey: Visitors' summer figures; means of arrival in Lake District*

|  | Bus | Train | Car | Coach | Other |
|---|---|---|---|---|---|
| **Men:** | | | | | |
| Under 21 | 2 | 9 | 4 | 2 | 1 |
| 21–34 | 14 | 16 | 4 | 1 | — |
| 35–54 | 12 | 18 | 10 | 6 | — |
| 55 and over | 10 | 14 | 3 | 3 | — |
| Total | 38 | 57 | 21 | 12 | 1 |
| **Women:** | | | | | |
| Under 21 | 6 | 6 | 8 | 4 | — |
| 21–34 | 14 | 24 | 15 | 4 | — |
| 35–54 | 20 | 32 | 10 | 5 | 3 |
| 55 and over | 21 | 23 | 9 | 9 | — |
| Total | 61 | 85 | 42 | 22 | 3 |
| **Total:*** | | | | | |
| Under 21 | 11 | 20 | 16 | 8 | 6 |
| 21–34 | 30 | 46 | 22 | 5 | — |
| 35–54 | 36 | 52 | 23 | 11 | 3 |
| 55 and over | 35 | 37 | 12 | 12 | — |
| Total | 112 | 155 | 73 | 36 | 9 |
| Percentage of total | 29 | 41 | 19 | 9 | 2 |
| Percentage in cross-section† | 11 | 17 | 61 | 10 | 1 |

\* Includes some for which sex is not specified.
† The cross-section of visitors interviewed throughout the Lake District.

duced almost identical results, this was not so with visitors and the tables are concerned only with the summer sample. Again, the most noticeable feature was the predominance of women: only 37 per cent of passengers were men.

As already stated, nearly all the visitors were staying in the Lake District. About as many of those interviewed were using a bus five or more times a week during their holiday as were doing so on one to four days a week, and presumably this means that the most frequent travellers are outnumbered by the second group by two or three to one. Some 41 per cent of visitors interviewed on buses had come to the Lake District by train. The 29 per cent stated to have come by bus is probably an inflated figure, as some of the people interviewed were actually on their way to the Lake District. This 29 per cent may be compared with the 11 per cent of our cross-section of holiday-makers who said they had travelled to the area by bus: see Table 17. The bottom two lines of Table 16 also show that while 61 per cent of the general sample of holiday-makers reached the Lake District by car, only 19 per cent of those interviewed on local buses did so. In fact this 19 per cent is a higher figure than anticipated, but there was other evidence of those who had made long-distance journeys by car using buses for local trips, sometimes travelling outward by one route, walking across country, and returning by another. These people included a few who had hitch-hiked to the area.

The winter sample of visitors using buses was small—eighty-one—and does not warrant detailed investigation. The winter bus users, however, generally seemed younger and included a higher proportion of men than the summer sample, while almost half of those interviewed had come to the Lake District by train.

### TOURIST TRAFFIC

We interviewed a cross-section of Lake District holiday-makers at points where they could be expected to have used a representative selection of means of travel. Subjects were picked at random, and all members of the family or party were then included in the survey. Visitors (those spending one or more nights in the Lake District) have been separated from day-trippers, and parties of ten or more people have been excluded from the tables.

*Means of Arrival: Tables 17 and 18*

Table 17 shows how visitors reached the Lake District and where they came from. Only 17 per cent had arrived in the area by train. The survey was late in the season, and a higher

## Tourist Traffic

figure would probably have been obtained in one of the Wakes Weeks of June and July. But taking other evidence (such as railway census figures) into account, it is clear that trains take a lower proportion of the traffic to the Lakes than is normal for holiday areas.[1]

Some 28 per cent of the visitors were young people travelling on their own. These accounted for 19 per cent of car arrivals, 45 per cent of train arrivals, 45 per cent of bus arrivals, and 32 per cent of coach arrivals. It was found that on average the young visitors had not travelled so far to reach the Lake District as had the others, which partly accounts for the greater use of buses. But a large proportion of the young people who had come long distances had travelled by train.

A further selection of visitors were interviewed on trains and at stations. All those who had travelled to the Lake District by train, and who were staying in a town served by a branch line, were asked if they would have visited the same place had the branch trains been withdrawn.

Regular visitors are less likely to be put off by the closure of the railway than are new-comers; but against this, younger people are less likely to be put off than older ones, being more able to take in their stride the change from train to bus. Many of those asked whether they would still have visited Keswick if the railway there had been closed said they would have chosen Windermere instead.

### Local Travel: Table 19

Other questions were asked to discover how the visitors were travelling within the Lake District during their holiday. The table includes separately the answers given by those interviewed on trains and stations. Of the total of 840 questioned off railway premises, only 45 said they were using trains during their holiday. Of these, 33 were using a train once or twice,[2] and only 8 were using the railway more frequently. None had, or intended buying, a runabout ticket. A number of visitors relying on public transport to see the Lake District had not

---

[1] See British Travel and Holidays Association figures.
[2] The inconsistency between this figure and the top line of the table is due to four car arrivals having been interviewed on a train.

## Research Work

### TABLE SEVENTEEN
*Visitors' survey: Cross-section sample of those staying in the Lake District*

| Arrived by | Number | Percentage | District of origin | Number | Percentage |
|---|---|---|---|---|---|
| Car | 508 | 60½ | Lancashire | 199 | 30 |
| Train | 141 | 17 | London and South East | 187 | 28 |
| Bus* | 96 | 11½ | Other North England | 141 | 21½ |
| Coach* | 85 | 10 | Midlands and East Anglia | 67 | 10 |
| Other | 10 | 1 | South and West | 19 | 3 |
| Total | 840 | 100 | Abroad | 20 | 3 |
| | | | Wales | 15 | 2½ |
| | | | Scotland | 14 | 2 |
| | | | Total | 662† | 100 |

\* Buses are stage-service vehicles calling at all stops on request. Coaches are long-distance express vehicles running to timetable but not serving local requirements. Here the term does not include vehicles on day excursions.

† The question about district of origin was omitted in some interviews.

### TABLE EIGHTEEN
*Lake District visitors' survey: Probable effects upon visitors of closing branch lines in the towns concerned*

| | Would have come | Would not have come | Age | Would have come | Would not have come |
|---|---|---|---|---|---|
| Have been to Lakes before | 134 | 44 | Over 25 | 122 | 54 |
| Have not been to Lakes before | 79 | 41 | Under 25* | 68 | 17 |
| Previous holidays unstated | 7 | 1 | Age unstated | 30 | 15 |
| Total | 220 | 86 | Total | 220 | 86 |

\* In this survey we recorded the age of only the leader of each family or party, and the under-25 group are young people on holiday without parents or other adults.

heard of runabout tickets. Several visitors questioned in Keswick did not even know that the town had a railway station. About one-third of the visitors used (or definitely intended to use) buses during their holidays, some outings involving both bus and walking. Nearly half the holiday-makers intended doing some walking, but only 1 per cent were cycling.

The sample of day trippers was fairly small—444. Of these, 271 had come by car, 91 by train, 30 by bus, and 52 by coach. Only 12 per cent of the day-trippers were young people not accompanied by adults, and most of these came by car. Almost exactly half the trippers who came by train said that they would still have come to the Lake District had all its railways been closed, but some would not have penetrated the area so deeply.

During the course of interviewing visitors and trippers we questioned a member of each of ten large parties, whose total membership was 294. Nine of the parties consisted of visitors and one of day-trippers. Four had come by train, four by coach, and one each by car and by bus.

TABLE NINETEEN

*Visitors' survey: Means used by visitors staying in the Lake District for travel during holiday*

|  | Car arrivals* | Non-car arrivals | |
|---|---|---|---|
|  |  | Interviewed off trains and stations | Interviewed on trains and stations |
| Train: |  |  |  |
|   One or two journeys | 13 | 24 | 48 |
|   More than two journeys | 4 | 4 | 19 |
|   Runabout ticket | — | — | 12 |
|   No answer | 3 | 6 | 7 |
| Bus | 69 | 228 | 203 |
| Coach | 21 | 55 | 48 |
| Boat | 165† | 130‡ | 71 |
| Car | 505 | 26 | 25 |
| Cycle | — | 13 | — |
| Walk | 158 | 202 | 159 |
| Numbers covered | 552 | 332 | 266 |

\* Including some interviewed on railway property.
† Includes twelve interviewed on boat.
‡ Includes eight interviewed on boat.

CHAPTER NINE

# CASE HISTORIES

### I. THE TEIGN VALLEY

THE story of public transport in most areas is long and complicated and many facets have usually to be ignored. But it is perhaps worth tracing developments step by step in one valley.

Numerous and bold were the plans for an inland railway between Exeter and Newton Abbot *via* the Teign Valley, but eventually it took two companies to complete the single track. The Teign Valley Railway (Heathfield–Christow) needed nine Acts of Parliament to coax it into existence and another three afterwards—a record for a line of seven and three-quarter miles. At one time the London & South Western Railway seemed likely to help, as part of a scheme to push into Great Western Railway territory and compete for Torquay traffic, but nothing came of this. Having been in and out of Chancery, the Teign Valley Company allowed their line to be built under G.W.R. protection and to be worked by the G.W.R. when completed, from Heathfield to Ashton, on 9 October 1882. Soon goods went another mile to Christow station. The line was built of standard gauge, while until 1892 the Newton Abbot–Heathfield–Moretonhampstead line was broad. The Teign Valley line was thus an isolated concern with locomotive and locomotive shed (at Ashton), rolling stock and *esprit de corps* all its own.

The Exeter Railway Company were sanctioned in 1883 to finish the route to St. Thomas, Exeter. They were saddled with a hopelessly large capital. Engineering works involved two tunnels and numerous cuttings. 600,000 cubic yards of earth had to be removed—largely necessary only to placate landowners. But at the opening in 1903 enthusiasm ran high, it being claimed that the new hinterland opened up would greatly extend Exeter's trade.

## 1. The Teign Valley

The complete St. Thomas–Heathfield run was known as the Teign Valley and worked as a single unit by the G.W.R., which bought out the two independent companies at a fraction of the cost of construction. In the 1920's there were six daily passenger trains and several new halts were added belatedly to meet road competition: before the coming of motors several hamlets beside the railway pleaded in vain for a station. The passenger service reached its peak in the mid-1930's, with nine daily services each way and extras on certain days: Chudleigh Knighton Halt, for example, was the start for a twice-weekly trip to Newton Abbot. Passengers at Christow and Trusham included mine and quarry workers, and the mines and quarries produced heavy mineral traffic.

The Teign Valley was a useful alternative to the main line by Dawlish and Teignmouth and during World War II thousands of pounds (£7,000 at Longdown alone) were spent on increasing its capacity, although only on a few occasions were services diverted this way. Immediately after the war the railway still held the monopoly of public passenger transport at Christow, Ashton and Trusham, and between these places and Exeter and Newton Abbot. But Christow Parish Council's request for the restoration of several pre-war trains was refused. The Devon General Omnibus Company were approached and began an Exeter–Christow service. The railway immediately lost much traffic on this section, though workmen's trains and the last train from Exeter on Saturday nights for which there was no bus alternative remained fairly popular. Already there were complaints that railway management was out of touch with local conditions: changes in working hours by the biggest employer in the valley were not accompanied by revised train times, and for many years a tea-time train ran virtually empty from Heathfield to Christow, the children for whom it was originally provided having transferred to coach travel.

While mineral traffic at Christow and Trusham brought in several thousand pounds a month, the branch may have covered its high costs; but little was done to save money where possible, and no effort was made to publicize the route as an attractive alternative to the main line for sightseers. (Even so, people travelling 'for the ride' often outnumbered local travellers.)

## Case Histories

After 1950, however, mineral traffic fell sharply at Christow, and as more people bought cars passenger business fell further. For several years before closure the only goods train ran from Heathfield to Christow and back once a day and it would have been possible to concentrate at Christow all the 'crossing' of trains on the single line, abolishing the other signal-boxes, at Trusham and Longdown. At Longdown the loop line was in fact removed so that trains could no longer pass there, but the unnecessary signal-box was still staffed two turns a day at a cost of about £1,300 a year.

A few months before closure British Railways did introduce a timetable which greatly reduced expenditure. Crossing movements were concentrated at Christow and had the line remained open this would have prepared the way for closing the two other signal-boxes. Trains now mainly ran between Exeter and Newton Abbot instead of merely from Exeter to Heathfield. The last train on Mondays to Fridays no longer waited nearly an hour and a half at Heathfield before returning almost empty to Exeter, but went on to Newton Abbot, enabling the branch to be open about ten hours less each week. The first up morning train now started from Newton Abbot instead of coming down the branch first. Had these and other economies been made in 1950, as they could well have been, the Transport Commission would have saved at least £30,000.

On 9 June 1958 passenger services were withdrawn and the St. Thomas–Christow section closed completely. In their submission to the Transport Users' Consultative Committee, British Railways said that on average the branch was used for 193 daily journeys and that closure would give a net saving of £15,125, allowing for the cost of subsidizing an alternative bus service for Ashton and Trusham. The figures were based on operation of the 'traditional' time-table, not the more rational one introduced at the eleventh hour.

A year after closure, when we interviewed the former regular travellers we also investigated the operation of the goods service, running three times weekly from Heathfield to Christow. Eleven men had been employed on track maintenance when all sixteen miles of the route were used daily by passenger trains; seven men were allocated to look after the eight miles remaining open

## 1. The Teign Valley

for the thrice-weekly freight train. Level-crossing keepers were being paid in full as before. As an 'economy', the loop at Trusham (where concrete traffic was heavy) had been put out of action, though not removed. This often meant that goods trains which could otherwise have terminated at Trusham had to go all the way to Christow (where goods business had almost ceased) merely to enable engines to change ends. Staff were employed at Ashton and Christow after traffic had virtually vanished. Eventually nature intervened: in 1960 floods washed out the track near Ashton; Trusham, its loop line reopened, became the terminus. Another curious economy was the demolition of the old engine-shed at Ashton by a gang of workers brought to the village daily: an offer to purchase it by the local coal merchant was rejected.

The valley's bus services hardly provide a happier story. For many years there has been an Exeter–Longdown–Leigh Cross–Moretonhampstead service, on which the buses leave the main road to call at Dunsford. The immediate post-war introduction of the Exeter–Leigh Cross–Christow service has also been mentioned. By the time the railway closed, occasional Christow buses were calling at Bridford and a Saturday-only Exeter–Leigh Cross–Dunsford service had been started. After much wrangling and examination of the state of the winding Teign Valley road, the Devon General were subsidized by British Railways to run from Leigh Cross (connecting with the other services there) to Newton Abbot.

The timetable was based on the 'traditional' railway one (i.e. not on the last-minute amendments) and repeated its expensive mistake of operating the whole service from the Exeter end. Not only would mileage have been saved had the bus been stationed at Newton Abbot, but the journey down from Exeter in the morning and back in the evening duplicated existing services between Exeter and Leigh Cross. With great care the service made it possible for quarry workers to reach Trusham by public transport although none had used the trains immediately before their withdrawal. Although most people leave work at 5.30 p.m., departures from Exeter (Christow service) and Newton Abbot were not until 6.25 and 6.10 respectively. Whereas conflicting movements on the main line made earlier

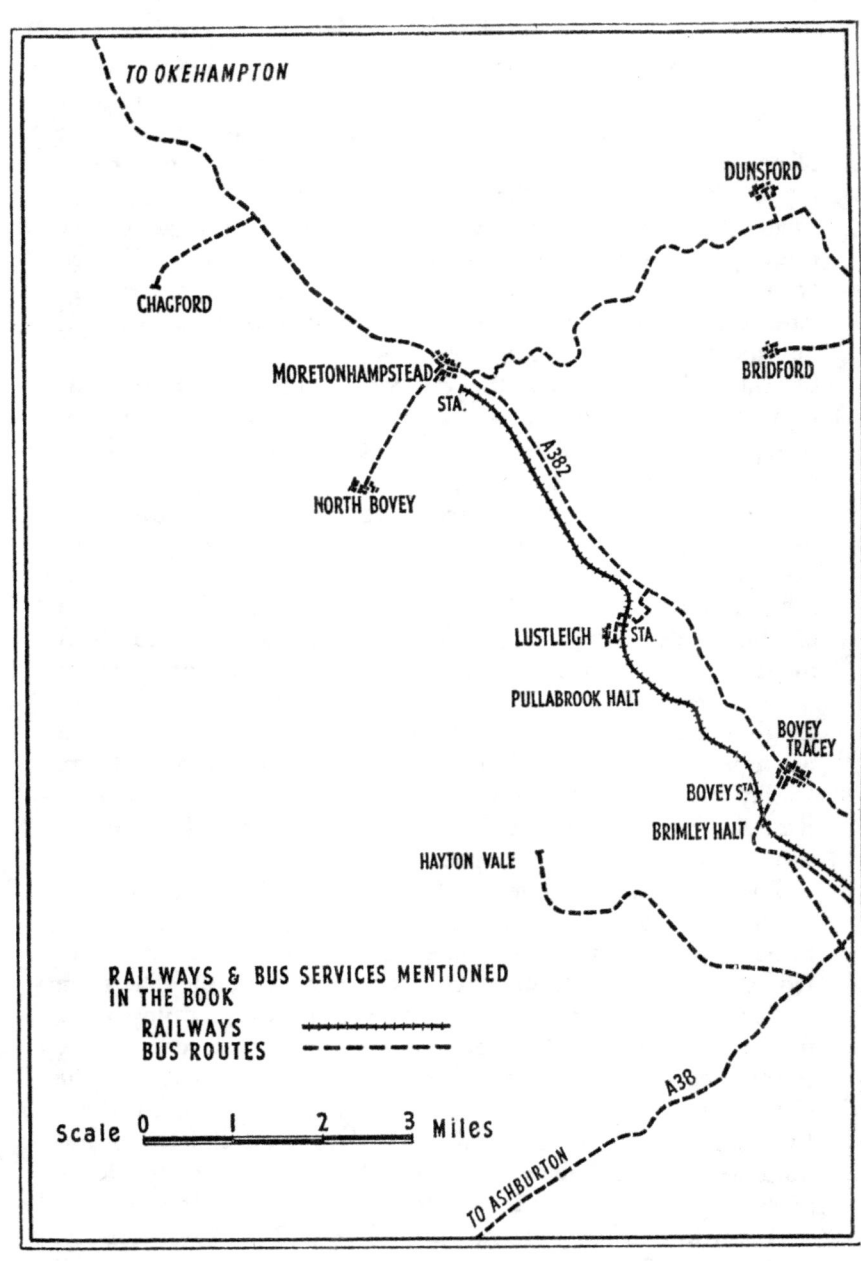

The Teign Valley and

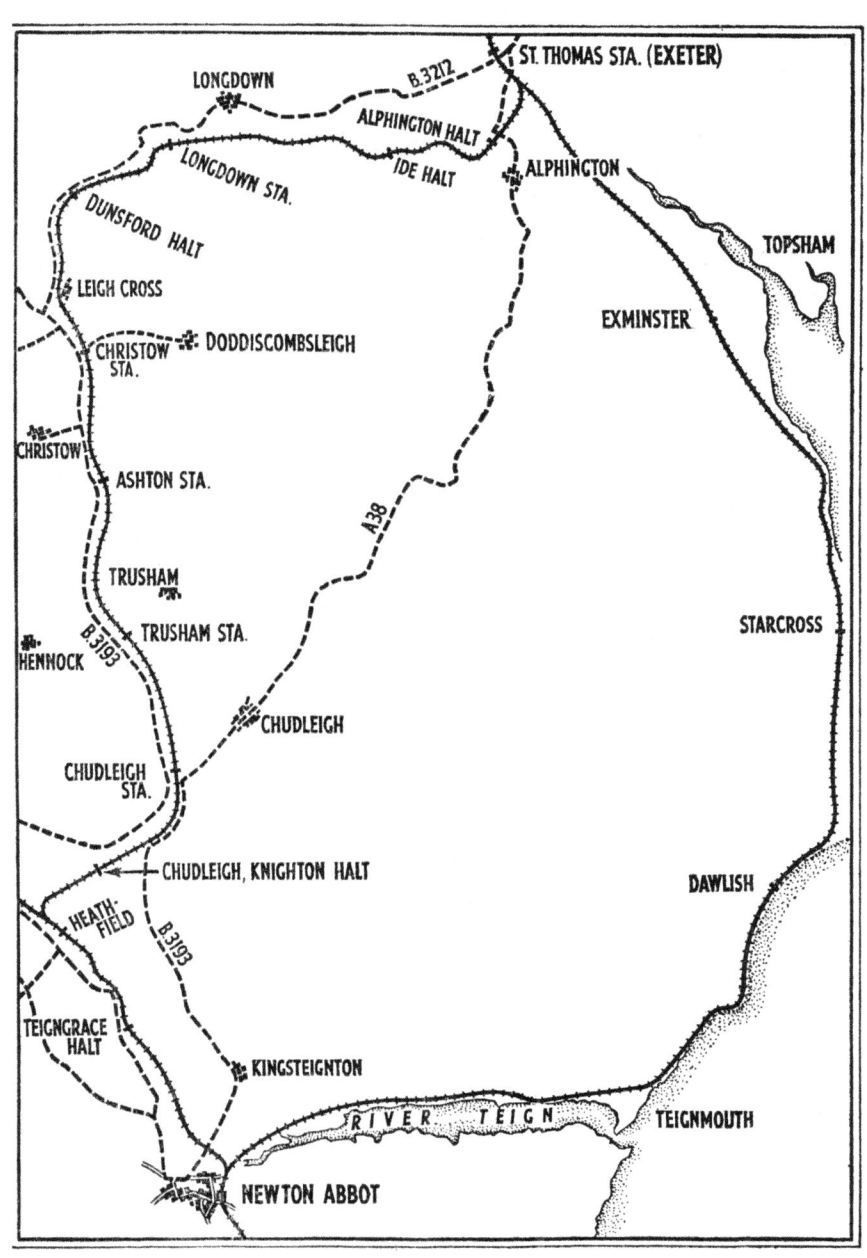

Moretonhampstead areas.

## Case Histories

railway departures difficult, this did not apply to the buses: at Newton Abbot the bus was idle between 5.44 p.m. and 6.10, for example. Moreover, the bus took much longer *en route* than did the trains, a punctual start thus being more necessary. Using the bus, Ashton girls working in Exeter would have been away from 7.49 a.m. to 7.33 p.m.: on the closure of the railway all four girl commuters immediately became scooter travellers.[1] Although the last down Saturday evening train had consistently been the most popular, no corresponding bus was provided and this alone made several people buy cars. Again, had the bus service been based on the railways' last-minute improved timetable, and had the vehicle been garaged at Newton Abbot instead of Exeter, this mistake might have been avoided.

The catalogue of errors is almost inexhaustible. Although subsidized by British Railways, the bus was not mentioned in the railways' timetable and did not run to the railway station (nearly a mile from the bus station) at Newton Abbot. But though the bus service was not run as part of the railway system, the Devon General refused to accept it as part of their bus network. Because it was subsidized, its vehicle was kept rigidly separate, though clearly a through Exeter–Newton Abbot service calling at Christow would have been far cheaper and more convenient. The best arrangement might have been for the Devon General to concentrate on a minimum through Exeter–Newton Abbot route with a few journeys *via* Christow, and for

[1] Later the 6.25 p.m. from Exeter to Christow was retimed to leave at 6.0 p.m., on the belated realization that many Christow people were leaving by the 5.45 p.m. Exeter–Leigh Cross—Moretonhampstead service and walking from Leigh Cross. This led to the underemployment of the 6.25 while the 6.0 often had to run in duplicate. The long-overdue adjustment helped Christow and Bridford, but no effort was made to bring the Teign Valley service into line—it continued running at the same times as before. People bound from Exeter to Ashton and Trusham thus left 25 minutes earlier but spent these 25 minutes waiting at Leigh Cross. As already said, the 6.10 p.m. from Newton Abbot could have left earlier—at least by 5.55 p.m.—and a quicker turnround at Leigh Cross (passengers were kept waiting six minutes even after the bus had arrived) would almost have eliminated the 25-minute gap. As British Railways fixed the timetable but otherwise took no interest and the Devon General ran the service on a contract basis, taking no initiative, it is likely that the 25-minute gap was never discussed.

## 1. *The Teign Valley*

this to be supplemented by a smaller, privately operated bus serving the villages—Dunsford, Bridford, Christow and possibly also Canon Teign, Teign Village, Ashton and Trusham with a market-day service—and making additional connections at Leigh Cross with the Exeter–Moretonhampstead service.

Such an arrangement would not, however, have appealed to the Devon General. Although the company treat fairly the existing small operators with whom they come in contact, they have always been anxious to keep new concerns from their territory. A small private operator did in fact seek a licence for a once-weekly Newton Abbot market service, taking in Doddiscombsleigh, Higher Ashton and Trusham. The application was opposed, partly on the grounds that there was no justification for a new operator and partly because of a by-law prohibiting vehicles of more than fifteen seats from using the road from the valley to Trusham. Not only did the small man lose his case, but as the result he had to discontinue a school bus he had run successfully over the same route for some years. Technically the road was closed to buses, but elsewhere a blind eye has been turned to the occasional stage bus (as distinct from coaches) and here this would have caused little inconvenience and greatly have strengthened the village's economy.

The Devon General believe that complete control of an area leads to the most economic working. Within limits this might be true, but it was hard to understand the company's attitude in this case. Even if the one-man concern could seriously have been regarded as a rival, the Traffic Commissioners more than adequately protect existing operators. Incidentally, the small operator alleged that the Devon General had taken steps to prevent him buying one of their unwanted vehicles, which was sold elsewhere.

It would be wrong, however, to portray the Devon General as grasping monopolists providing the minimum of services. Even in the Teign Valley they have shown generosity. Though taking no steps to make the best of the subsidized service, for which the whole responsibility was left with British Railways, on their own territory they have provided village services on a scale which cannot possibly pay. The Christow service itself is liberal, and the journeys routed *via* Bridford have made a great

## Case Histories

difference to life in the moorland village.[1] Even Dunsford has a service twice on Saturdays. Because the Christow bus can scarcely call at Bridford and Dunsford *en route*, and because the subsidized railway service is strictly segregated, this involves running separate vehicles all the way down from Exeter.[2]

Obviously there may be different views as to how the valley's public transport could be organized most efficiently. But there can be no doubt that a far more useful service could consistently have been provided with less strain on the resources of the Devon General and British Railways. The subsidized bus service in particular could better have met the valley's needs with half the mileage had the important journeys been correctly timed.

On-the-spot investigations in the valley by responsible officials have been lacking. Had an official of the Devon General been based for a fortnight on Leigh Cross at any time during recent years, he might have been able to make recommendations resulting in considerable savings. He might even have discovered such facts as that over a considerable period seven Bridford workers returning home by lorry from Exeter were dropped at the main road just after the evening bus had started its climb to the village. Not only did the bus miss these passengers by a narrow margin, but on the return journey it was frequently delayed when meeting the taxi they had hired. These workers, and others from Christow, used to be employed in the mine at Christow and are now builders in Exeter: with initiative it might even have been possible to reorganize the bus service so that they used it the whole way from Exeter.

Certainly a lengthy visit from an official would have increased

---

[1] At first Christow passengers (the majority) were subjected to trips up and down the lane to Bridford both on their way to work in Exeter in the morning and going home at night. After complaints, sensibly Christow passengers travel *via* Bridford in one direction and Bridford people *via* Christow in the other.

[2] Until the 6.25 p.m. Exeter–Christow was changed to 6.0, on Saturday evenings three buses left Leigh Cross travelling westward simultaneously. On 11 July 1959 the order was: 1, Teign Valley route, with two passengers for Ashton only; 2, Bridford *via* Christow, with eight for Christow and four for Bridford; 3, Doddiscombsleigh, with fourteen passengers. Bus 2 returned to Exeter with seven passengers and bus 3 with four.

## 2. Newton Abbot–Moretonhampstead

goodwill. The bus company have had a poor reputation in the valley: because they have been so palpably out of touch with local conditions, villagers have assumed (perhaps unfairly) that it is useless making suggestions for improvements. Instead of seeking changes in the bus service, the population has set out to depend on it as little as possible. When we finished our investigations, employers were allowing their workpeople to return home in the firms' lorries and vans; a considerable number of journeys were made in privately-owned or works' mini-buses; more people were probably given lifts in other people's cars than travelled by bus; many preferred to stay at home as much as possible and thus be able to afford a taxi when they had to go away; because regular buses were shunned wherever possible, there was a lively demand for organized coach excursions; throughout the valley the motor-cycle ownership rate was high. But for all this, the buses still fulfilled some essential requirements.

### 2. NEWTON ABBOT–MORETONHAMPSTEAD

The Newton Abbot–Moretonhampstead branch, twelve miles long and rising 550 feet, was built by the independent Moretonhampstead & South Devon Railway in 1866. It was worked by the main-line company, the South Devon Railway, which however contributed only £5,000 to the capital of £105,000, and provided a poor initial train service. Traffic was sparse until the 1890s, when well-filled trains of six or seven six-wheeled coaches became familiar at the Moretonhampstead terminus with its wooden overall roof and mean exterior. Then, in 1906, the Great Western Railway, which by now had absorbed both the South Devon and the Moretonhampstead Companies, began running buses to Chagford in connection with the trains. As the Torbay resorts grew, visitors flocked to the railway.

The train service developed more as the result of road competition than of the increase in traffic. In 1910, when the railway still had a virtual monopoly of rapid travel, there were only five daily trains each way on most days; but in the summer of 1935, by which time most people on this route used the road,

## Case Histories

eleven trains ran from Newton Abbot to Moretonhampstead and a further six to Bovey Tracey. An express non-stop from Torquay to Bovey Tracey and through trains from Exeter to Bovey Tracey were among the expedients used to combat road competition at various times. Two new halts were opened—they would of course have had greater value before the advent of motors—and at Heathfield (junction with the Teign Valley line) the station was enlarged and the signal-box provided with apparatus more advanced than was found almost anywhere on the main line before 1950.

During the 1939–45 war, and for some years afterwards, eight daily trains ran each way. A ninth was added about 1950 at the request of Moretonhampstead Parish Council, but a little-used afternoon train was later withdrawn: the timetable in 1958 is printed below. (Although a train started at Moretonhampstead in the morning and finished there in the evening, its locomotive spent the night at Newton Abbot; it was thought cheaper to incur the 'light' mileage than to retain the small engine-shed at Moretonhampstead.)

With only eight daily trains, the railway could not hope to compete with the frequent bus service between Newton Abbot and Bovey Tracey. After the war Bovey Tracey thus provided relatively few passengers. But the railway still gave the best-

**Table 90   NEWTON ABBOT, BOVEY and MORETONHAMPSTEAD**
(Second class only, except where otherwise shown)
WEEK DAYS ONLY

| Miles | Station | am | am A | pm | pm | pm | pm | pm |
|---|---|---|---|---|---|---|---|---|
| 0 | Newton Abbot dep | 7 50 | 9 20 | 1250 | 2 15 | 4 25 | 6 5 | 8 15 |
| 2 | Teigngrace Halt | 7 54 | 9 24 | 1254 | 2 19 | 4 28 | 6 8 | 8 19 |
| 3 | Heathfield | 8 0 | 9 29 | 1258 | 2 23 | 4 33 | 6 13 | 8 23 |
| 5 | Brimley Halt | 8 4 | 9 33 | 1 2 | 2 27 | 4 37 | 6 17 | 8 29 |
| 6 | Bovey | 8 10 | 9 36 | 1 6 | 2 30 | 4 40 | 6 20 | 8 31 |
| 7 | Pullabrook Halt | 8 15 | 9 41 | 1 11 | 2 35 | 4 45 | 6 25 | 8 36 |
| 8 | Lustleigh | 8 19 | 9 45 | 1 15 | 2 39 | 4 50 | 6 29 | 8 40 |
| 12 | Moretonhampstead arr | 8 31 | 9 57 | 1 26 | 2 51 | 5 1 | 6 41 | 8 52 |

| Miles | Station | am | am | am A | pm | pm | pm | pm |
|---|---|---|---|---|---|---|---|---|
| 0 | Moretonhampstead dep | 7 50 | 8 40 | 1015 | 1 35 | 3 15 | 5 10 | 7 6 |
| 3 | Lustleigh | 7 59 | 8 48 | 1023 | 1 43 | 3 23 | 5 18 | 7 — |
| 4 | Pullabrook Halt | 8 1 | 8 50 | 1025 | 1 45 | 3 25 | 5 20 | 7 10 |
| 6 | Bovey | 8 8 | 8 54 | 1031 | 1 50 | 3 30 | 5 25 | 7 16 |
| 6 | Brimley Halt | 8 11 | 8 58 | 1034 | 1 53 | 3 32 | 5 27 | 7 19 |
| 8 | Heathfield | 8 16 | 9 3 | 1044 | 1 58 | 3 39 | 5 34 | 7 24 |
| 10 | Teigngrace Halt | 8 20 | 9 7 | 1048 | 2 2 | 3 43 | 5 37 | 7 28 |
| 12 | Newton Abbot arr | 8 25 | 9 12 | 1053 | 2 9 | 3 50 | 5 42 | 7 34 |

A  First and Second class

A Road Motor Service is operated by the Devon General Omnibus Company between Moretonhampstead & Chagford.

## 2. Newton Abbot–Moretonhampstead

timed and fastest service for Lustleigh and Moretonhampstead. Tourist traffic was still considerable, although little effort was made to attract it.

When economies became essential, the trains chiefly used by tourists were selected as victims. All Sunday trains and the one daily through service from Paignton and Torquay ceased running on 30 June 1958, though they were of course included in the timetable for the whole summer and were just about to carry their peak loads. The economy exercise seemed particularly pointless since before another summer had dawned the whole line had closed. And British Railways failed to profit from the cuts even during the line's short remaining life. For example, the 8.40 a.m. from Moretonhampstead used to work through to Paignton, but after the cuts it terminated at Newton Abbot. Train and engine were thus available to work the next train back from Newton Abbot to Moretonhampstead. But a different engine and train (prepared considerably in advance) were still supplied: passengers and staff alike complained of the waste. At the end, with daily passenger-train mileage a mere 170, locomotives were allocated for sixteen hours.

Even without a major recasting of the timetable, other changes could have been made more usefully. Early in the morning, two trains were employed on the branch, arriving at Newton Abbot at 8.25 and 9.12, whereas a single service arriving at 8.45 would have saved half the running costs and would almost certainly have won more passengers than did the two together. Most commuters were girls beginning work at 9.0, and 8.45 would also have given better main-line connections. Evening return services for workers were equally unsatisfactory, with no departure from Newton Abbot between 4.25 and 6.5. Out of forty commuters questioned on the 6.5 one evening, thirty-two said that to leave at least twenty minutes earlier would have suited them better. Commuters from Moretonhampstead had to leave at 7.50 a.m. and did not return there until 6.41 p.m., even though they worked in Newton Abbot only between 9.0 a.m. and 5.0 or 5.30 p.m. Here was a typical example of a rural commuter service which had remained unaltered while working hours shortened. If two trains were ever needed on the branch, it was during the early

## Case Histories

evening; but one diesel car could have handled the entire service.

The real need, however, was for an entirely new timetable providing trains for specialist requirements only. In the days when most local people used the trains for most purposes, a service running throughout the day, summer and winter, might have been justified. But after the war a high proportion of the traffic fell into well-defined categories, and many of the trains were nearly empty. For example, three signal-boxes had to be kept open an additional two hours for the last train in one direction only which frequently had no passengers and except on Saturdays seldom more than fifteen. Most commuters and long-distance passengers could have been carried by a single service each way in the morning and evening without any 'light' journeys. Two additional trains each way on Wednesdays (Newton Abbot market day) and three on Saturdays would have covered 80 per cent of winter needs—a total of sixteen trains a week compared with the forty-two plus six light journeys run even after the 1958 cuts. In summer a minimum of five daily trains was perhaps needed, including the commuter services and through services from Paignton and Torquay for holiday-makers.

|                                                                                           | £     |
|-------------------------------------------------------------------------------------------|-------|
| Cost of operation                                                                         | 6,445 |
| Additional staff costs to meet essential signalling requirements above those needed for freight working | 1,951 |
| Additional cartage costs[1]                                                               | 1,530 |
|                                                                                           | 9,926 |
| Less present revenue (1958)                                                               | 4,628 |
| Estimated loss                                                                            | 5,298 |

('Therefore an 80 per cent increase of passengers would be necessary for a lightweight diesel to cover its cost of operation.')

In presenting their case for closure, British Railways claimed that the net annual saving would be £17,319. This figure was the result of deducting the estimated annual loss of traffic receipts of about £4,600 and additional road cartage costs of about £1,500 from the gross saving of £23,500. It was alleged

[1] Road conveyance of parcels which could not be taken on a small diesel unit.

## 2. Newton Abbot–Moretonhampstead

that about a further £1,100 could be saved by the withdrawal of passenger trains on the Ashburton branch, it being stated that the Totnes–Ashburton services 'are integrated to some extent with those operating on the Newton Abbot–Moretonhampstead line', a curious claim entirely new to those who heard it. British Railways' submission admitted that diesel traction would lower costs: with diesels the expenses of a passenger service were estimated as shown on the previous page.

This estimate covered purchasing a lightweight diesel unit with inadequate parcels accommodation for use exclusively on the branch, with a timetable presumably similar to that in operation. But as we have seen, the need was not for a self-contained fairly frequent service but for a skeleton one with through trains to and from the Torbay branch. Although this would have necessitated a bigger unit with a higher cost per mile, the annual mileage would have been less than half and the following advantages would have accrued:

1. In summer through trains to and from Torquay and Paignton would have had far greater traffic potential than a self-contained branch service.

2. A larger unit could have carried parcels, saving the £1,500 required for parcel lorries.

3. With a skeleton diesel service the goods train could have had sole occupation of the branch above Bovey Tracey and Moretonhampstead signal-box could have closed, saving over £1,000 annually.

There would seem to have been a good chance of making the passenger services positively pay, though it would have been strongly in British Railways' interests to retain them even at a small loss. There were two reasons for this. Firstly, many people who used partly the branch and partly the main line or Torbay line gave up all their rail travel, and the loss of revenue on these other lines must have amounted to at least £1,000 and probably £1,500 annually.[1] Secondly, the branch provided the only

---

[1] This is our estimate, based partly on interviewing regular travellers, partly on reports from hotels and boarding-houses in the areas served, and partly on the declining sales of runabout tickets: British Railways did not reveal what allowance if any had been made in the estimated loss of gross receipts (£4,628) for loss of fares elsewhere.

## Case Histories

convenient rail access to the moors from Torbay, the West Country's leading holiday area; its closure has made holidays by rail in South Devon less attractive. As several other branches in the area made more serious losses and their demise was inevitable, every possible effort should have been made to keep Moretonhampstead on the system.

The South Western Transport Users' Consultative Committee in fact recommended that the service should be retained, while the Central Committee, though advising immediate closure, said that reopening should be considered when diesel power became available in 1961. Not content with that, a deputation of local people went to see the Parliamentary Secretary to the Minister of Transport. Despite their emphasis that management of the branch had been poor, a circular letter despatched by the Ministry nearly three months later merely stressed that 'the Western Region of British Railways has every interest in making the right decisions'. The Ministry, as the Transport Users' Consultative Committees, accepted British Railways' estimates without question.

Although the Central Committee recommended that reopening be considered, soon after the closure the Press quoted a railway official as saying that it was most unlikely that trains would be restored. In fact reopening has not been considered. When questioned, a spokesman at Paddington stated that the allocation of diesel units was on a 'planned basis' and as none had been planned for Moretonhampstead, reopening was impossible.[1]

Supposing the passenger service had to be closed, had they cared to be enterprising British Railways would have run at least one or two experimental excursions from Paignton and Torquay. But no train was allowed even for Moretonhampstead

[1] More recently it seems that British Railways would like to close the Bovey Tracey–Moretonhampstead section even to goods traffic, but that they are unwilling to present a case for this to the Transport Users' Consultative Committee in case the question of restoring passenger trains might be raised. Goods traffic remains heavy on the first part of the route, and since 1961 has been augmented by bananas taken to a new ripening plant beside Heathfield station. The siding to this depot leads off from the Teign Valley branch: although the truncated Teign Valley line is itself now virtually a siding, special signals have been installed to govern the new point.

## 2. Newton Abbot–Moretonhampstead

carnival which had previously attracted several hundred rail travellers each year; the line was 'closed' and comparisons with Welsh lines closed to regular traffic but still used by excursions were of no avail. Applications for two guaranteed hired trains were greeted coolly, but accepted after long deliberation.

The track is now maintained only for goods and technically is subject to a 15 m.p.h. speed limit; but in fact it is in such good order that a train can comfortably keep pace with the bus winding along the road above Bovey Tracey.

Had the bus service materially benefited from the closure, the picture would be brighter. But it has not. The results of our interviews with regular train travellers have already been described. Probably less than one in twenty of the journeys made by holiday-makers on trains have been transferred to the bus service. Whereas the trains succeeded in grouping together several classes of business (local travel by local people, beginnings and ends of long-distance journeys, day trips by holiday-makers, and conveyance of parcels), the bus is largely concerned with the purely local transport of residents. Fewer long-distance travellers now visit the area by public transport, and of those who do a large proportion complete their journey from Newton Abbot by taxi. Most trippers relying on public services now use excursion coaches from the resorts: though some of these coaches are run by a subsidiary of the Devon General, the traffic does not contribute to the upkeep of the route's stage service. Parcels are still carried by British Railways, presumably still at a cost of over £1,500 annually.

Much damage might have been avoided had certain buses been provided with room for luggage, made station connections at Newton Abbot and had a place in the railway timetable. As it was, Bovey Tracey and Lustleigh were more or less wiped off the nation's transport system, and Moretonhampstead was only slightly luckier: there is a long-established bus service from Moretonhampstead direct to Exeter, where passengers alight near the long-distance coach station and the Southern Region station, but no connections are advertised. And although the Newton Abbot–Moretonhampstead and the Exeter–Plymouth buses shared the same road for a short distance near Heathfield

*Case Histories*

until 1961, they made no connections with each other and not even a shelter was provided. Now the Exeter–Plymouth service is routed *via* Bovey Tracey, which thus also has a direct link with Exeter: this move would of course have been more valuable taken less than two years after the railway's closure.

If regarded as a self-contained unit, the Newton Abbot–Moretonhampstead–Okehampton service is run efficiently and offers less to criticize than the Teign Valley. Commuter services, however, leave something to be desired. True, the Devon General's 8.58 a.m. arrival in Newton Abbot is far better than the railways' 9.12, but no evening bus leaves Newton Abbot between 5.5 and 6.10, giving the majority of commuters a long wait. This is particularly undesirable in view of the length of the journey. The bus takes twice as long as did the trains to and from Moretonhampstead: the road above Bovey Tracey is perhaps the worst in Devon in relation to the vehicles carried, and the bus makes three diversions *en route*, to serve Heathfield, the Brimley area of Bovey Tracey and Lustleigh.

Running a commuter bus each way cutting out Heathfield and Brimley (which have other services) would have helped. At one time buses ran through from Newton Abbot to Okehampton, but owing to thin traffic at the Okehampton end single-deckers now work this part and passengers change at Chagford. A better idea, at least at certain times, might have been to run the double-decker between Newton Abbot and Lustleigh, passengers for Moretonhampstead and beyond changing into the single-decker on the main road at its Lustleigh junction and thus avoiding having to travel *via* Lustleigh itself. But as in the Teign Valley the subsidy complication arises: no doubt it has been thought best to leave well alone the arrangement by which British Railways pay for the Lustleigh call.

One last point which illustrates the attitude of big companies to one-man concerns. When the trains were withdrawn, the Devon General's service did not include a Sunday morning trip suitable for taking Bovey Tracey people to the sea, and a request for one was rejected. The independent operator mentioned in the Teign Valley 'case history' was approached and agreed to apply for a licence; but the Devon General then at

142

*3. Coniston*

short notice put on the required bus. It would have been wrong for the small operator to have been allowed to steal traffic during hours that the Devon General were running, but as said before, the licensing authority very adequately protects the interests of established operators. The point is that while the Devon General may well have been commercially right in refusing to run the service in the first instance, the small man with his personal touch might possibly have made it successful.

### 3. CONISTON

Coniston in the Lake District lost its passenger trains in October 1958. The story of events leading up to the closure, and of the aftermath, is one of the most complicated and controversial yet encountered by the author in the study of rural transport.

The Coniston branch, nine and three-quarter miles long, left the Coastal or Furness line at Foxfield, was single track, included heavy gradients, and commanded some fine Lakeland scenery. It carried fair local traffic all the year round, and summer passengers included long-distance travellers on their way to or from a Lake District holiday and numerous day trippers. It had eight daily trains, plus an additional service between Foxfield and Broughton, the most important intermediate station. In summer the regular services were supplemented by twice-weekly excursions from Blackpool and Morecambe.

The local authorities, the Friends of the Lake District, and the general public, including a number of well-known people, produced evidence which persuaded many members of the North Western Transport Users' Consultative Committee that this was a case where the withdrawal of trains might cause widespread inconvenience.

The majority of the independent members asked that British Railways should continue the service. But the votes of British Railways' own representatives gave a majority in favour of endorsing the closure proposal.[1] It was, however, agreed that

[1] Following criticisms of the handling of this case, instructions were given that British Railways' representatives should cease to vote at meetings of the Consultative Committees.

## Case Histories

trains should continue until a satisfactory bus alternative had been provided. The complication here was that the Lancashire County Council were seeking an order prohibiting vehicles of more than fifty hundredweight laden weight from using the Coniston–Foxfield road.

These were the statistics submitted to the North Western Transport Users' Consultative Committee:

*Figures presented by British Railways in connection with the withdrawal of passenger trains between Foxfield and Coniston*

|  | Average number of passengers joining and alighting per day ||||
| --- | --- | --- | --- | --- |
|  | *Broughton* | *Woodland* | *Torver* | *Coniston* |
| Winter—Mondays to Fridays: | | | | |
| Joining | 89 | 16 | 41 | 80 |
| Alighting | 73 | 14 | 41 | 78 |
| Summer—Mondays to Fridays: | | | | |
| Joining | 111 | 12 | 9 | 86 |
| Alighting | 76 | 12 | 36 | 138 |
| Winter—Saturdays: | | | | |
| Joining | 94 | 13 | 21 | 66 |
| Alighting | 113 | 14 | 18 | 81 |
| Summer—Saturdays: | | | | |
| Joining | 127 | 29 | 29 | 137 |
| Alighting | 173 | 43 | 40 | 133 |

*Minimum gross estimated savings resulting from the proposed withdrawal*

|  | £ | £ |
| --- | --- | --- |
| (a) Minimum gross estimated savings per annum |  | 19,435 |
| (b) Estimated loss of traffic receipts from passenger traffic | 2,756 |  |
| (c) Estimated additional road cartage costs | Nil |  |
| (d) Minimum net annual economy, resulting from proposal |  | 16,679 |
| (e) Additional savings, not assessed, will accrue from maintenance and renewal on station buildings |  | — |

## 3. Coniston

The Central Consultative Committee supported the recommendation that the line should be closed, but asked the North Western Committee to withdraw the condition about the bus service. At the same time the Central Committee gave a broad general hint to guide the North Western Committee: consequential expenditure by local authorities and other organizations should not be offset against the savings claimed by the Transport Commission. The saving to the railway was sufficient to justify a decision to recommend withdrawal of the service.

The North Western Committee did not accept the first directive, and reaffirmed their recommendation that trains should be withdrawn only when a suitable bus service was introduced. Further deadlock was avoided by British Railways agreeing to pay Ribble Motor Services to run the bus service, which was allowed to use the road upon certain conditions laid down by the Traffic Commissioners. Briefly, the restriction sought by the County Council on large vehicles was approved, but an exception was made for stage-service buses provided a number of passing places were built.

While the railway was running, the route of course had no bus. Ribble already operated their Ambleside–Coniston and Ulverston–Coniston routes at a loss, and felt disinclined to start another obviously uneconomic venture. They introduced the new service purely as agent for the railways, who have paid about £5,000 a year—receiving back only about £2,000 in fares.

The number of bus journeys provided was little more than half the number of trains, in winter the service dropping to four daily trips each way, with an additional trip—paid for by the Lancashire County Council—for school children during term time. During the Lake District Transport Inquiry, it was found that only a small proportion of the former train travellers were using the bus. Three classes of people were dependent on it:

1. Long-distance passengers without access to a car, who could not afford a taxi and had luggage too heavy to carry from the bus to the railway station at Ulverston. Some of these people said they would prefer to travel by Ulverston if the bus called at the railway station there, or by Windermere if there were a through Coniston–Windermere bus.

## Case Histories

2. Workers, shoppers and others travelling from places formerly served by the branch line to places on the main Coastal line. For those bound for places north of Foxfield, the alternative bus services from Coniston were no use, and though Barrow could be reached *via* Ulverston, this was indirect. When the railway was working twelve Coniston people used to travel daily to Barrow; in 1961 only three. Several of the others were interviewed and complained that the journey by bus was too long and expensive to make it worth earning the higher wages obtainable in Barrow.

3. Residents along the bus route going to another place on the same route. These purely local passengers, not changing into trains at Foxfield, accounted for over half the traffic.

All three classes of passengers complained bitterly about the fares. The return from Coniston to Foxfield was 3s. 1d. for about twenty miles, compared with only 2s. 3d. from Coniston to Ambleside. To go to Barrow *via* Ulverston was cheaper than to use the more direct Foxfield route. There were no through road-rail bus fares.

The bus was paid for by one organization and run by another. Local people alleged that because of the ease of passing the buck it was impossible to come to grips with either. They believed that British Railways were satisfied to see the traffic decline as this might hasten the day when the subsidy could be withdrawn. Certainly not even transport officials expected the arrangement to last indefinitely: Ribble Motors stated that they would not run the bus without a subsidy.

Not only was the bus lamentably failing to pay its way, but the Inquiry team were left with the impression that a subsidy of £3,000 could be used more fruitfully elsewhere. Yet even here the money could have been better spent. Quite apart from *advertised* misconnections, lost connections were frequent. The timetable failed to allow for the almost standard late running of some trains on the Furness route. And although the service was subsidized by British Railways, it was not mentioned in the railway timetable, so few visitors could know of it.

It might be argued that a transport operator who withdraws or intends to withdraw a service should allow a reasonable time for the public to adjust to the changed conditions. The bus

## 3. Coniston

was presumably intended as this stopgap. But in practice everyone who could make alternative arrangements did so, leaving the bus with mere oddments of traffic.

Two years since our main investigations the position remains substantially the same, although British Railways cannot be expected to continue the subsidy much longer.

As suggested in the *Lake District Transport Report*, it would be possible to reduce the service substantially and still cater for the essential local requirements on a minimum basis, while routing one daily Coniston–Ulverston bus each way *via* Ulverston station would greatly ease the problem of long-distance connections. But no matter how many cuts are made on the Foxfield service, it is highly unlikely to pay its costs, and although the Ulverston station diversion might be in the long-term interests of public transport as a whole, it would probably involve Ribble in a loss. Though much waste might have been avoided, we are still left with the question: who pays?

The Coniston–Foxfield service received national notoriety in January 1962 when in bad weather a school bus left the road and turned on its side in a field. Three children were seriously injured and four slightly injured. At the request of the parents and of Lancashire County Council, for a time the school bus was diverted to a much longer route (admitted to be no less dangerous) and the question of stopping the subsidized service was also raised. No permanent changes have been effected, but the accident stressed the need for further improvements to the Class III road between Foxfield and Torver.

Passing places had to be built before the bus service could begin, and by 1961 the County Council had spent £13,000 on improving the route, while it was stated that eventually £40,000 would be required 'to put the road into something like a reasonable and safe condition'. The work already done has been at the expense of other schemes previously regarded as more urgent, and a County Council spokesman added: 'It would, we feel, be true to say that expenditure of the order of £40,000 would never have been contemplated on this road had not the rail service been withdrawn, although some minor improvements would have been made.'

The closure of the railway has necessitated heavy road

## Case Histories

expenditure, the basic bus service between Coniston and Foxfield loses £3,000 each year, and the inquiry into the school bus accident elicited the information that it was costing between £800 and £1,000 more a year to send children to school by road than by train. The Post Office and many businesses in Coniston have been put to extra expense, as have numerous individuals. The cessation of the trains has seriously lessened Coniston's tourist trade.

The closure of the railway has resulted in greater total expenditure and loss than its retention need have incurred, given a diesel unit and simplified staff and signalling arrangements. The annual loss on the passenger side could have been reduced to £5,000 a year while goods traffic continued. The Lake District Transport Inquiry could not, however, recommend reopening. Most of the regular daily passenger traffic was lost for good. Reopening would inevitably have had to be experimental, making local people reluctant again to become dependent on the service. Indeed, once the decision to withdraw passenger trains had been taken, complete closure would have been wiser—although it was churlish to refuse to run the occasional special excursion to Coniston while the line was still used by goods.

Retaining the branch for goods alone proved extravagant. The engineering department concerned claimed that cessation of passenger services allowed relatively little saving in track maintenance, and in fact shortly after they were withdrawn part of the track was relaid. Compared with the standards of several other regions, notably the Eastern, the upkeep of the Coniston branch was needlessly elaborate and costly for three light goods trains a week.

Because of the loss on the goods service, at the end of 1961 British Railways proposed complete closure of the branch, and goods trains ceased in spring 1962 after the Consultative Committee had intimated agreement. Opposite is a summary of the statistics backing British Railways' proposal.

It will be noted that staff costs alone amounted to almost twice the annual revenue.

Even that is not quite the end of the story. To handle traffic formerly dealt with at Broughton, British Railways have estab-

## 3. Coniston

lished a small goods depot at Foxfield, until then solely a passenger station. Although the capital cost of this work has not been stated, it must have been considerable. The railways are now planning to close nearly all small goods stations everywhere, concentrating traffic at a few large centres: almost certainly the new goods station at Foxfield will close by the end of the decade. In the meantime it might have proved cheaper to retain the branch as far as Broughton goods depot, or to transfer goods to another station on the Furness line.

*Figures presented by British Railways in connection with the withdrawal of freight traffic between Foxfield and Coniston*
*Minimum gross estimated savings resulting from the proposed withdrawal*

|  | £ | £ | £ |
|---|---|---|---|
| (a) Minimum gross estimated savings per annum | | | |
| (1) Staff costs | 4,629 | | |
| (2) Repair of rolling stock | 220 | | |
| (3) Train movement costs, other than staff | 301 | | |
| (4) Day-to-day costs in the repair of permanent way, bridges, buildings, signalling, roads, fences, etc., and day-to-day costs of working stations and signals, other than staff | 277 | | |
| | | 5,427 | |
| (b) Estimated loss of traffic receipts from traffic | | | |
| (1) Freight | 2,879 | | |
| (2) Miscellaneous | 362 | | |
| | | 3,241 | |
| (c) Estimated additional road cartage costs | | Nil | |
| (d) Minimum net annual economy, resulting from proposal | | | 2,186 |
| (e) The net estimated savings do not include any portion of the undernoted renewals expected to be required in the four years ending in 1964 | | | |
| In year 1961 | 5,940 | | |
| In years 1962 to 1964 | 12,080 | | |
| | 18,020 | | |

*Case Histories*

#### 4. THE KESWICK LINE

The study of the cross-country line from Penrith through Keswick to Workington was an important feature of the Lake District Transport Inquiry. It revealed much unnecessary expenditure over a considerable number of years. British Railways largely accepted our findings, which were given substantial publicity. Yet at the time of writing, eighteen months after their publication, practically no economies have been introduced.

This is what we said about the line in the *Lake District Transport Report*. Tenses have not been altered, and train service details refer to those in force in 1960.[1]

\* \* \*

Most of the survey work was undertaken in summer 1960, when British Railways were known to be considering the possibility of withdrawing passenger trains. In October an interim report of our findings was presented to the North Western Transport Users' Consultative Committee. Almost simultaneously, British Railways announced a reprieve of passenger services 'at least for the time being', and a conference of the Press and representatives of local organizations was later called by railway officials at Penrith. At this meeting, and also at an earlier meeting of the Inquiry's Steering Committee, British Railways largely agreed with the findings of the interim report, which estimated an annual loss of £50,000 on the line.

The railway is $39\frac{3}{4}$ miles long. (As a useful comparison, Workington to Penrith is farther than Paddington to Reading, and Keswick to Carlisle just as far.) About $13\frac{1}{2}$ miles (Penruddock to Threlkeld, and Brigham to Derwent Junction, near Workington, where the branch joins the Coastal line) is double track, a legacy of the days when substantial freight traffic used the route on its way from the east to the west coast. Gradients are severe, with several sections at 1 in 70, and a bank of four miles at 1 in 63 between Threlkeld and Troutbeck.

[1] Details of the mechanics of our investigations, perhaps the most thorough yet made on any branch line, and tables analysing traffic censuses were also included in *Lake District Transport Report* but are omitted here.

## 4. The Keswick Line

In the past the western end of the branch had substantial local passenger traffic: Brigham and 'Cockermouth were both junctions for other passenger services, and between Workington and Cockermouth several of the intermediate stations now closed were once important enough for trains to terminate and start there. Although the last purely local train between Workington and Cockermouth survived the war, today nearly all local passengers between these two places go by road, and passenger traffic is heavier at the eastern end of the line, between Keswick and Penrith. But all weekday trains travel right through from Penrith to Workington and vice versa.

All the branch lines which used to connect with this railway are closed even to goods, and there is no remnant of the once-substantial coal traffic at the western end. But freight is still heavier here than at the eastern end, Cockermouth being the distribution point for a range of goods, including oil, over a wide area.

The line was among the first in Britain to benefit from the introduction of diesel multiple-unit trains. These arrived in February 1955, when the service was appreciably increased and accelerated. Blencow station, which had previously closed, was reopened and remains open. But another intermediate station served by the diesels for several years has since closed: Embleton, between Bassenthwaite Lake and Cockermouth. There are now nine intermediate stations between Penrith and Workington, of which Keswick and Cockermouth are obviously the most important.

Another improvement introduced with dieselization was the running of a number of through trains to and from Carlisle (*via* Penrith). Also, since 1955 the line, which for many years had run on weekdays only, has had a Sunday service during the summer. Connections at Workington for Whitehaven have generally worsened since dieselization, however.

The basic weekday train service consists of eight diesels in each direction. During the summer, though not for the whole duration of the summer timetable, there is in addition a steam-hauled section of the *Lakes Express*, with through coaches to and from London; and on summer Saturdays a through service runs

## Case Histories

to and from Manchester. On summer Sundays there is a skeleton service of diesels and a steam train—the only one of the week with refreshment facilities—from Newcastle; on about half the Sundays during the season a special excursion runs from such places as Morecambe and Blackpool. On the freight side, daily trains run from Penrith to Keswick and back, and from Workington to Keswick and back. There is an additional early morning freight and parcels train from Penrith to Keswick: the locomotive works back light to Blencow, where it shunts in the limestone-quarry sidings.

We estimated that on Mondays to Fridays in September the total number of passengers at all intermediate stations, joining or alighting from the eight trains in each direction, averaged about 350 a day, and in winter little if any above 200 a day, or less than $1\frac{1}{2}$ per stop. On many days the average would be less than one person per booked stop. Though certain trains could run non-stop from Penrith to Keswick without inconveniencing anyone, other services draw a high proportion of their small total of passengers from intermediate stations.

In view of the paucity of traffic, the question is: what inconvenience would the withdrawal of trains cause?

*Local patronage.* The majority of local passengers would be able to travel by road without much hardship, though many journeys would take longer and cost more. Careful examination of the traffic suggested that if the railway closed only about one-third of the journeys would be made by bus instead of train. One-third would be made by private motor vehicle and the remaining third would cease. Visits to Carlisle would be especially reduced.

*Long-distance travellers.* Less than 20 per cent of Keswick's visitors arrive by train; of those who now do so, a considerable proportion would probably go elsewhere (mainly still within the Lake District) upon the closure of the branch. It was noted here as elsewhere in the Lake District that young people travelling long distances still mainly use trains. It is well known that older people return to scenes of earlier holidays at least occasionally. If, lacking a railway, fewer new visitors are introduced to Keswick and district, it is reasonable to suppose that the supply of 'regulars' and 'semi-regulars' in future years will

## 4. The Keswick Line

also suffer. As an incidental, the withdrawal of trains would almost certainly end the Keswick Convention.

From the point of view of the local economy, the railway's long-distance traffic may well be of greater importance than its scale suggests, though of course it might be a matter of indifference to British Railways that the closure of the branch now would result in Keswick receiving fewer motoring visitors in years to come.

The local population would also miss the railway connections with long-distance trains at Penrith and Carlisle, and cars and taxis would have to be used more often. The Transport Commission's present policy is against running a special service of buses for railway needs, and the adaptation of the existing bus service to give main-line connections at Penrith would not be satisfactory.

*Excursionists.* In summer the majority of passengers are apt to be holiday-makers on day excursions. The elimination of one or more possible trips from the Lakeland itinerary would cause them no hardship, but the line is part of the Lakeland scene and visitors would regret its loss: it is now used by many on 'round trips' up the Shap line to Penrith and down the Coastal line *via* Workington and Barrow. But most holiday-makers actually staying in Keswick use trains only for short journeys, and in the whole of August 1960 only sixty-four runabout tickets were sold at Keswick.

*Parcels and freight traffic.* Keswick's isolation would preclude any economic road service from equalling the present good passenger-train deliveries of newspapers and parcels. The withdrawal of freight facilities would, however, cause relatively little inconvenience. Deliveries of some goods would be delayed, road congestion would be slightly worse, and the price of coal might rise, but industries would not suffer seriously. Complete closure of the branch between Cockermouth and Blencow could be economically better justified than closure to passengers only. Penrith–Blencow would have to be kept for limestone traffic, and there would be a strong case for retaining Workington–Cockermouth for general freight. This would leave $12\frac{3}{4}$ miles of the present $39\frac{3}{4}$ miles.

## Case Histories

If this line serves definite passenger needs, with good parcels and freight facilities, how could the results be improved? What economies could be made?

*Train service.* Some trains are almost always well used while others, particularly at midday and at the western end of the branch, are nearly empty, summer and winter alike. It is suggested that the standard weekday service could be reduced to seven trains in each direction between Penrith and Keswick and to four or five beyond. A summary of suggestions for bringing the service more into line with traffic demands has been submitted to British Railways.

*Singling of track.* Even with the present service there is no justification for the retention of the $13\frac{1}{2}$ miles of double track.

*Signalling.* Again, even supposing the present train service was maintained, signalling costs could be materially cut. The suggestions below take into account the proposed reduction in train service. Five signal-boxes could be closed, saving at least £7,500 a year. *Brigham:* to become 'crossing' station between two single-track sections. There is an important level crossing here. *Cockermouth:* signal-box to be closed and only one platform of passenger station to be used. Access to goods yards at passenger station to be by use of key in the single-line electric token apparatus. This would repeat the arrangement already introduced at Cockermouth main goods yard some distance away. *Embleton:* signal-box to be closed and level crossing to become the responsibility of crossing-keeper living in station house. *Bassenthwaite Lake:* signal-box to be maintained as at present; there is a level crossing here. The shortening of the loop line would enable signalman to command all movements himself without the assistance of a porter. *Braithwaite:* signal-box to be abolished as intermediate block post, and level crossing to become a porter's responsibility. *Keswick:* number of signal-boxes to be reduced from two to one. *Threlkeld:* at present the track becomes double here; when it is singled a section from Keswick to Troutbeck is proposed, Threlkeld box being closed. *Troutbeck:* this would become a crossing station, and the signal-box, now open only when required, would be regularly operated. *Penruddock:* at present the double track becomes single here. It is suggested that Troutbeck-Blencow

## 4. The Keswick Line

should be one section when track is singled, and the signal-box here closed. *Blencow:* to remain as at present.

Some signalling economies would have been possible years ago. The signal-box at Cockermouth Junction was abolished only recently, but could well have been closed before the war. Embleton station was closed in 1958, but all trains still have virtually to stop there to exchange the electric token with the signalman, much to the annoyance of the local people who claim that they have been inconvenienced while little, if any, expense has been saved: a signal-box is not needed, and British Railways now intend appointing a crossing-keeper instead. At Keswick, expenditure of a few hundred pounds would have enabled one of the two signal-boxes to have closed twenty-five or more years ago.

*Unstaffed Halts.* Retaining staff for platform duties at most of the small stations seems entirely unjustified. On the other hand, some trains regularly pick up or set down workers, schoolchildren or shoppers at all stations: while the stations are there and the diesels, capable of rapid acceleration, are passing them it would seem foolish to prohibit such passengers from travelling, or indeed to inflict the loss of traffic upon the railway itself. It is strongly urged that most of the small stations should be converted into unstaffed halts, and that the number of trains stopping at them should be reduced in the light of experience. Probably about £4,000 annually could be saved by reducing station staffs.

The suggestion most frequently encountered was that traffic could easily be vastly increased. We do not accept this. Populations are small, and the average length of even local journeys is necessarily long, with substantial fares. An increasing number of people will travel by car whatever the train service, and the western end of the branch faces competition from a frequent bus service.

But certainly a number of things could be done. As against the economies that have been suggested, a small expenditure might be considered on improving connections at Workington for Whitehaven and the Coastal line. For some years it has been near impossible to travel from Keswick to Ravenglass for the Eskdale Railway and back in time for hotel dinner; running

## Case Histories

the second down train of the day rather earlier would provide a valuable connection. Then, at present the *Lakes Express* runs three weeks less than the summer timetable period, but runs on the three Saturdays concerned: to retain it for the remaining fifteen days would usefully simplify the timetable and cost comparatively little.

Publicity could be improved. Most Keswick hotels had details of coach but not of railway tours. In January 1961 the only railway notice-board in the town (at the bus station) displayed a torn timetable. Many local people travel by train to Carlisle on market day, the trip being cheap and quick, but no effort is made to advertise the train's advantage to visitors.

A good case can be made for retaining the line, primarily for the benefit of the local economy, though also for the well-being of British Railways themselves.[1] Closure would result in much through traffic taking to the road the whole way, and fewer runabout tickets would be sold. Keswick is the capital of the Northern Lakes, and to withdraw trains would, to a much greater extent than most closures, give the impression that the railways are dying.

It is estimated that the proposals made above could result in cutting the annual loss from about £50,000 to about £25,000. The loss would remain considerable, but there would be the satisfaction that the money was being better spent.

Yet a rapid 'spring clean' seems improbable. Permission has to be obtained from regional headquarters at Euston, which in turn is controlled by the Transport Commission and the Minister of Transport, to spend the small amount of capital needed to effect the annual savings, and everywhere British Railways are short of money for immediate use.

\* \* \*

The forecast in the last paragraph proved correct. In October 1962 the double-track sections and signalling arrangements were unchanged. The only significant alteration was in the train service. At the end of the Inquiry specific suggestions for reducing this were put to British Railways. It was felt that the standard service of eight daily trains throughout the route could

[1] But see page 123.

## 5. Mid-Northumberland Bus Operators

be reduced to seven between Penrith and Keswick, and five on to Workington. All Workington trains would have run through from Penrith, with no 'stand over' time in Keswick.

Instead, the 1960–61 winter timetable treated the route in two sections: about half the trains started or terminated at Keswick. Daily mileage was slightly cut, but the operating cost was greatly increased because diesel units and staff were kept idle at Keswick during a large part of the day. Local staff were highly critical of the arrangements.[1] In 1962 further revisions were made, and the service was then brought more or less into line with the suggestions made by the Inquiry.

### 5. MID-NORTHUMBERLAND BUS OPERATORS

A study was made of the services of twelve independent bus operators in Mid-Northumberland, most of them based on Hexham. Most of the firms were the same as those questioned in 1957 by a team directed by Northumberland Rural Community Council,[2] and Mr. Douglas Mennear, a member of that team, has kept in touch with them and provided the present author with regular information. The author has also visited them himself. Nearly all have allowed their annual returns to the Ministry of Transport to be copied, and some have also shown their balance sheets and other records.

It had been hoped to publish comparative figures for all

---

[1] In April 1962 the general manager of the London Midland Region refused facilities for the author to revisit the line to check the effect of the changes on the traffic pattern. The author was specifically requested not to attempt to discuss the question with the local staff, although these men were mainly well known to him. A request for free travel was refused on the grounds that the further inquiry was 'a private venture'; but it was stated that he would be welcome to discuss the line with the divisional manager at Barrow-in-Furness. However, in a letter dated 2 October 1962, exactly two years after the Lake District Inquiry issued its interim report on this line and British Railways announced a reprieve for passenger services, the divisional manager stated: 'There are several inquiries still going on concerning the C.K. & P. line, and I think it best not to say anything at this stage concerning the small savings that have already been made.'

[2] See *The Rural Transport Problem in Mid-Northumberland*, 1957, and the shorter *Northumberland Country Bus*, 1958, both published by the Rural Community Council.

## Case Histories

twelve over a ten-year period, but financial and legal difficulties and deaths make it difficult to mention some firms by name, while amalgamations render some comparisons awkward. The table shows details of the decline in the business of four fairly typical firms.

The fleets of the twelve operators range from a single bus, generally used only twice weekly, to over twenty vehicles. Some firms are almost wholly committed to stage-service work, while others are primarily engaged in school and some industrial contracts and regard timetable services as a sideline. In some cases bus operation is combined with running a garage or other business.

Dividing total costs equally among total bus mileage (stage, contract and private hire), all firms have for some years made a loss on stage services. Some of the smaller firms have consistently run at a total loss, their fleets and equipment steadily being run down. Most have pruned mileage as well as raising fares, but with a few temporary exceptions the benefits have been more than offset by decline in traffic.

The chief object of this investigation was to gauge the advantages and disadvantages of a group of independent operators, rather than a combine, serving a rural area. The only services run by a combine in the survey area were of an inter-urban nature on the main roads.

We found that in every case the independents have a lower operating cost than the combines serving similar territory. Most firms estimated their costs at between 1s. 5d. and 1s. 10d. a mile, including depreciation and overheads; one sizeable operator was in 1962 still running at 1s. a mile. The use of free family labour and differences in accountancy methods, especially for depreciation and allocation of overheads between bus and any other business run, make precise comparisons impossible. It is clear, however, that a combine would spend something between a quarter and a third as much again on running the same mileage.

Most of the services have been carefully adjusted to meet changing demands. The population is scattered and the traffic potential poor, so that a request for a diversion to serve even two or three houses is taken seriously. 'Every detail counts',

## 5. Mid-Northumberland Bus Operators

### TABLE TWENTY

#### (I)

|      | Passengers | Miles   | Receipts (£) |
|------|------------|---------|--------------|
| 1953 | 95,945     | 61,621  | 3,310        |
| 1954 | 90,552     | 61,546  | 3,263        |
| 1955 | 86,223     | 61,942  | 3,240        |
| 1956 | 74,899     | 58,274  | 2,900        |
| 1957 | 80,655     | 58,550  | 3,455        |
| 1958 | 72,211     | 58,427  | 3,332        |
| 1959 | 59,837     | 79,093* | 3,083        |
| 1960 | 53,634     | 73,387  | 2,859        |
| 1961 | 66,283†    | 73,143  | 3,009        |

\* One service re-routed with increased mileage.

† Trading estate opened on route. A town bus service has since been extended to the estate and the above operator's passenger traffic is now well below the 1960 level.

#### (II)

|      | Passengers | Miles  | Receipts (£) |
|------|------------|--------|--------------|
| 1953 | 89,999     | 63,411 | 2,910        |
| 1954 | 78,916     | 63,160 | 2,691        |
| 1955 | 82,616     | 60,865 | 2,910        |
| 1956 | 80,006     | 60,865 | 2,852        |
| 1957 | 66,378     | 51,579 | 2,420        |
| 1958 | 78,740     | 41,084 | 2,487        |
| 1959 | 64,492     | 39,048 | 2,733        |
| 1960 | 70,128     | 39,406 | 2,817        |
| 1961 | 69,499     | 37,484 | 2,696        |

#### (III)

|      | Passengers | Miles    | Receipts (£) |
|------|------------|----------|--------------|
| 1953 | 176,068    | 131,836  | 5,349        |
| 1954 | 173,195    | 130,432  | 5,074        |
| 1955 | 184,279*   | 169,676* | 6,231*       |
| 1956 | 181,280    | 169,540  | 6,366        |
| 1957 | 180,179    | 160,564  | 7,276        |
| 1958 | 178,321    | 157,848  | 7,048        |
| 1959 | 165,179    | 153,012  | 6,588        |
| 1960 | 153,187    | 151,370  | 6,281        |
| 1961 | 142,975    | 151,370  | 5,875        |
| 1962 |            | 140,000  | 5,466        |

\* Additional service added.

## Case Histories

(IV)

| | Passengers | Miles | Receipts (£) |
|---|---|---|---|
| 1955 | 183,518 | 136,685 | 6,342 |
| 1956 | 177,305 | 127,425 | 5,933 |
| 1957 | 198,623 | 114,664 | 5,448 |
| 1958 | 180,586 | 115,056 | 6,317 |
| 1959 | 121,964* | 117,954 | 6,051 |
| 1960 | 85,536* | 113,510 | 5,854 |
| 1961 | 86,286 | 102,836 | 5,421 |

* Decline partly accounted for by run-down in hospital on route. Some school children have been lost to a new firm specializing in private hire.

said several of the operators. Their local knowledge and ability to improvise rapidly is their strongest asset. No combine with centralized control could possibly take such individual initiative to acquire passengers.[1]

Because of the sparsity of traffic, the independent operators assiduously meet each other's services even if not on speaking terms. The routes of the rivals thus form a closer-knit system than will generally be found in districts under the single management of one combine.

There seemed to be two drawbacks to services being provided by the independent operators. Firstly, few of them have up-to-date timetables, and those who have are unable to display them prominently in Hexham, Newcastle-upon-Tyne and other towns. This causes some inconvenience, especially to people travelling into Mid-Northumberland from other parts of the

[1] That well-defined heavy traffic flows are few may be gauged from the fact that of 227 passengers questioned on one of the more important routes on a Saturday (morning, afternoon and evening), only fifty-seven said they used the same bus as frequently as once a week; many used it between once a fortnight and once a quarter. Timings on this route were arranged with utmost care to suit people on the largest possible variety of journeys. Tight connections were made with a number of other services, and fifty-six of the passengers took advantage of these. Had connections been less efficient, a third of the people transferring from one bus to another might well not have made their journeys at all, and that would have been equivalent to a loss of nearly 10 per cent of the total traffic. Ten per cent of the traffic could easily have been lost in each case if schedules less exactly matched the requirements of those going home from work at lunch time, visiting relatives at a chest hospital, and going to Hexham for evening entertainment.

## 5. Mid-Northumberland Bus Operators

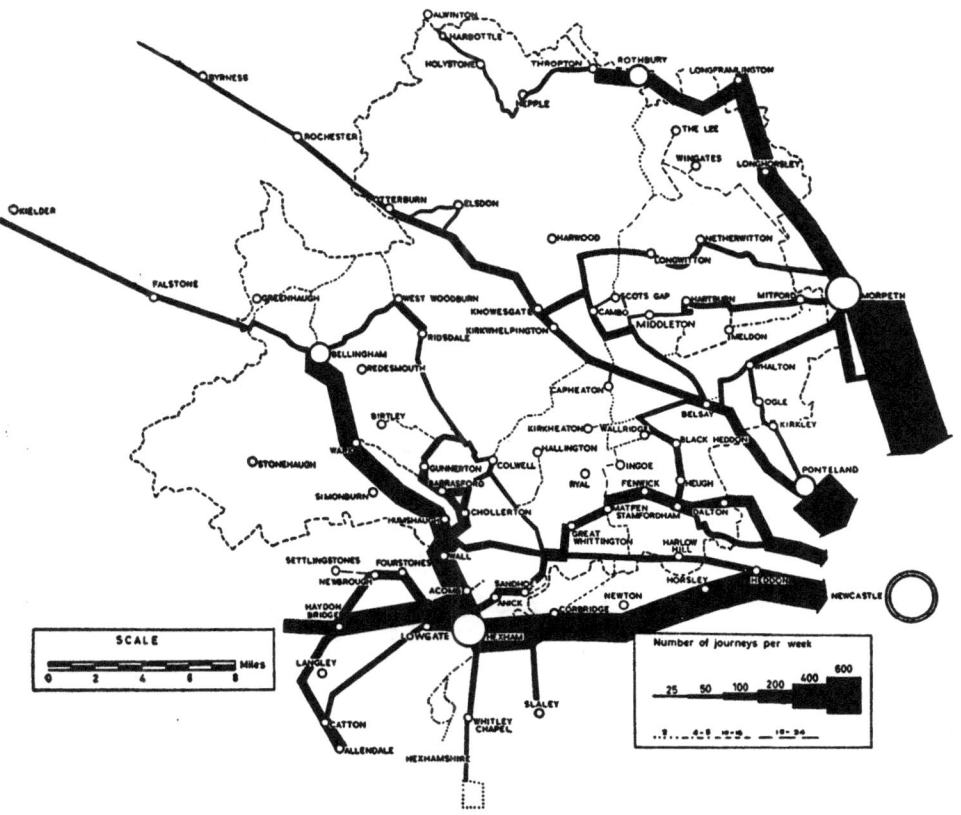

Mid-Northumberland Rural Transport Survey Area. Frequency of Bus Services, 1951

country, and some casual business might have been lost, especially with tourists. Then, the small firms offer a lower standard of comfort than the combines provide. Although modern coaches kept for excursion work are sometimes put 'on service', passengers generally travel in oldish vehicles showing definite signs of wear.

Taking pros and cons together, the small men served the district more efficiently and cheaply than would a large company, and the local population spoke highly of most of the services. Significantly there were no complaints about the relative roughness of some vehicles.

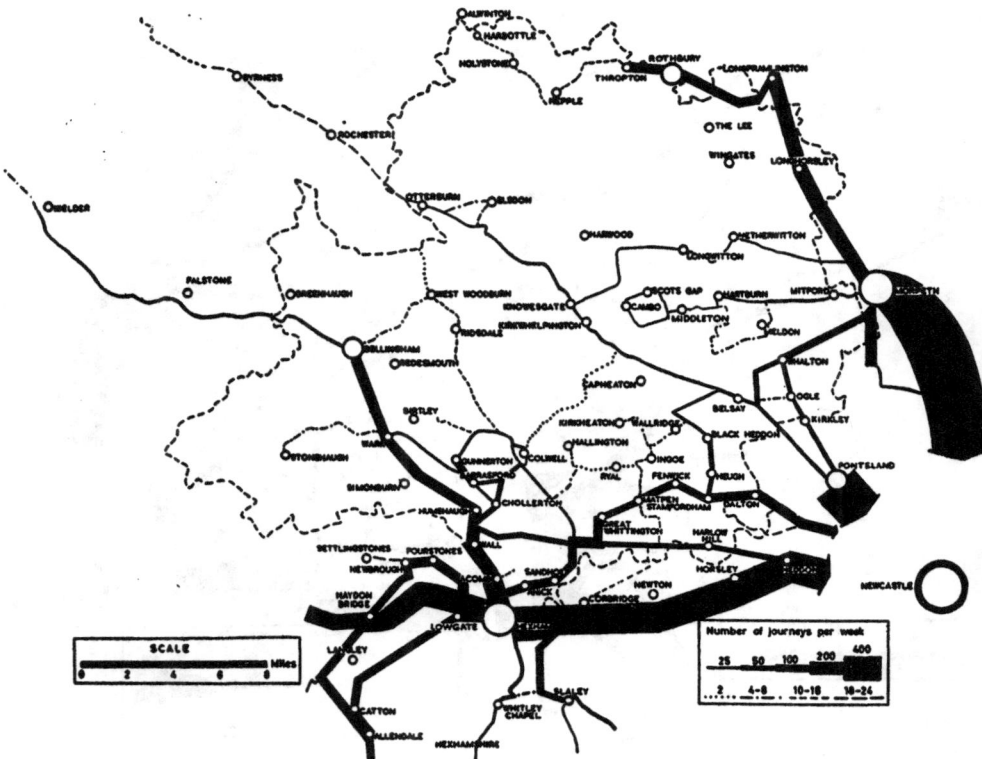

Mid-Northumberland Rural Transport Survey Area. Frequency of Bus Services, 1960

The small operators have many natural advantages. It seemed to us, however, that they were not fully taking advantage of their position, and that they could have run even more efficiently and cheaply. For example, overlapping between operators caused some wasteful mileage; one small operator, whose business consistently makes a total loss, virtually only creams off some of the market-day traffic from other routes. Methods of administration vary greatly. Thus, in 1960, while one man kept elaborate statistics and considered adjustments to his services almost weekly, another decided to go all out for maximum gross revenue without detailed costing, and a third

## 5. Mid-Northumberland Bus Operators

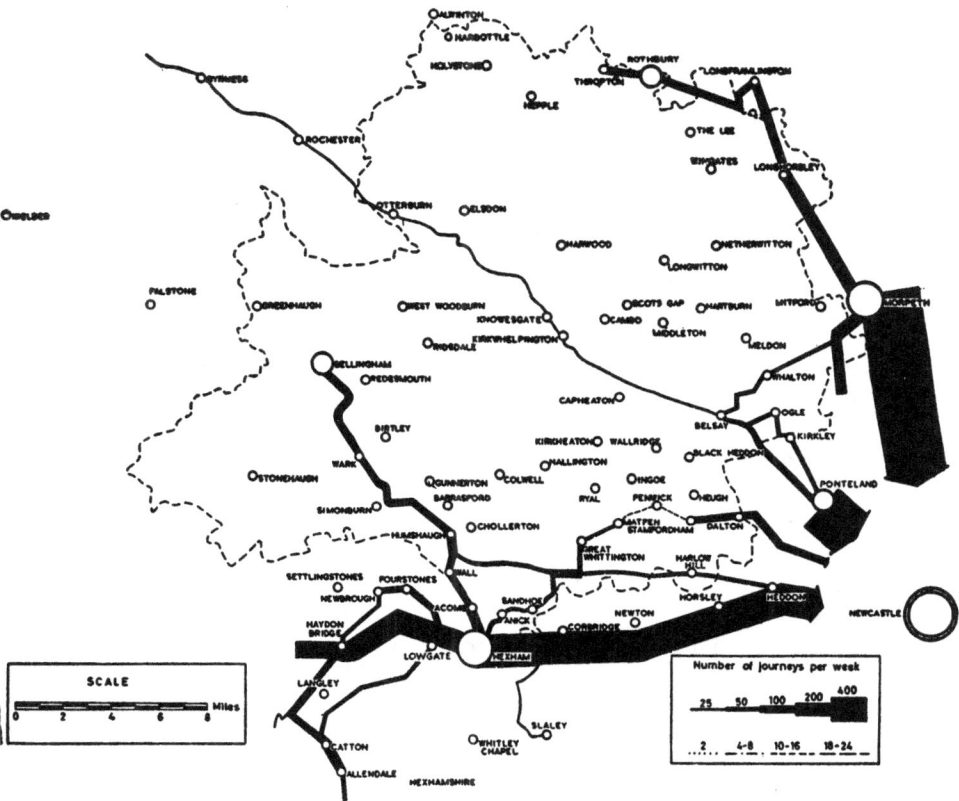

Mid-Northumberland Rural Transport Survey Area. Frequency of Bus Services, forecast for 1965

decided to 'strike the happy mean' by reducing mileage faster than the fall in passengers.

Several firms have clearly been wasting resources, but there is nobody to point this out—nobody to whom they can turn for advice. Were a qualified representative of the Traffic Commissioners able to allocate a tenth of his time to the welfare of these twelve independent operators, he might be able to hammer out changes reducing the total costs by a third without loss of traffic.

The Traffic Commissioners could also help immensely as

## Case Histories

intermediary between independents and combine, in this case United Automobile Services. Could the independents' services be included in United's timetable, for instance? Might it not be to everybody's benefit if one or two of United's routes were sold to independents? And could the sale of second-hand buses by the combine to the independents be eased? (It took high-level intervention, initiated by a voluntary organization, to obtain a suitable second-hand bus to keep one operator going.)

There were among the other points emerging from the investigations:

1. A number of operators remain in business even when incurring a steady loss, the value of their fleets steadily sinking. One or two of them would fail if creditors closed in, and one or two others will soon reach this position. The operators are subjected to great pressure from their customers and neighbours to keep services going—the man mentioned in brackets at the end of the last paragraph reversed his decision to give up at a village meeting of high fervour. For years of course there has lurked the hope that Government aid lay round the corner. And some drift on because they have no other employment to which to turn. Reductions in fleets and staff have brought a temporary lull in losses in some cases; in others—where for example a vehicle is no longer spare for profitable contract work on Saturdays—the situation has rapidly deteriorated further.

In official quarters the view persists that many of the losses in rural transport are made only on paper, and that operators would quickly retire if they ceased to cover themselves. This is just not true. (Elsewhere I have met men who have had to sell two old buses to pay the hire-purchase instalments on a more modern one, and had to dismiss staff to conserve funds for cash-down payment on fuel oil.)

2. Combining bus services with running a garage or smallholding seems to work well, enabling the bus side to have more staff at peak than at slack periods, and usefully spreading overheads.

3. We found clear-cut examples of operators with stage-service liability losing contract work—including school contracts—to firms concentrating on this more profitable business.

## 5. Mid-Northumberland Bus Operators

One school contract was taken from a firm with stage services for the sake of a few shillings between tenders. If this action led to the firm withdrawing its stage services, the education authority would have had to hire a special school bus on another route. The cost would be three or four times the amount spent on children's fares.

4. Despite the overall shortage of traffic, mini-buses could not be profitably used in this area, and one operator regrets having bought one. Generally every driver is needed on a larger vehicle during the morning rush hour, and it is cheaper to use a 29-seater at slack times than a mini-bus kept specially for the purpose. Where a mini-bus can be used in addition to the rest of the fleet on contract work during the rush hour, it would of course be useful for stage services on quiet mornings and evenings—provided a larger vehicle were available if needed.

5. Losses are made even by many services which appear bustling with activity at certain times. Not only are daily traffic peaks extremely sharp, but prolonged bad weather and harvest periods discourage business and severely reduce average takings. Thus a bus which normally carries twenty people may be used only by four or five during snow, gales and heavy rain, and also at hay and corn harvest. So the average throughout the year may be only thirteen or fourteen. This frequently makes the difference between profit and loss.

# POSTSCRIPT

### THE BEECHING PLAN

The publication of *The Reshaping of British Railways*[1] (the Beeching plan) has been the most significant event for rural transport in the few months since the rest of this book was written. The plan of course calls for drastic surgery, including the closure of over 2,000 stations and the virtual cessation of local passenger business in country areas. Once more little understanding is shown of the rural situation, and the following three points call for particular comment.

1. The report sets out with the preconceived idea that branch lines and stopping passenger trains on main lines must go. Its sole arguments about them are to justify this assumption. Thus a table setting out examples of the 'annual financial effect of withdrawing typical passenger services' includes a heavy vertical line demonstrating the point 'at which the withdrawal of services becomes financially justified'. No attempt is made to examine methods of operation, timetables or staffing arrangements, to establish whether services could not be run more cheaply. Dr. Beeching's aim is to shape a Victorian system to today's needs; but from the examples of the costings given, it is assumed that certain cheaper Continental methods must be permanently denied to Britain, that Victorian signalling systems shall be maintained to the end, and that stations require traditionally elaborate staffing and other arrangements. The average annual cost of a station, for instance, is put at £2,500. The report does not mention unstaffed halts, although these figure prominently in everybody's suggestions for running branch lines except those of British Railways.[2]

---

[1] Stationery Office, 1963.

[2] The Central Transport Users' Consultative Committee have been pressing for the introduction of more unstaffed halts. 'The British Transport Commission, on the other hand, have put before us a paper criticizing the

*Postscript*

Similarly no attempt is made seriously to assess whether careful pruning of train services and raising of fares in borderline cases might not achieve valuable results. These matters are dismissed with a few phrases of rhetoric, although from the figures in the report itself it is possible to demonstrate that fewer trains and higher fares could make a line pay with a substantially smaller traffic flow.[1]

2. Even supposing that branch lines if retained would continue to be run in the present costly manner, some of the forecasts in the report seem biased. For example, the theoretic minimum number of weekly passengers said to be necessary to make a line pay is worked out on the basis of higher costings than those given in the examples of the actual financial savings from the closure of specific lines.[2]

3. The report makes the classic mistake of assuming that because bus services run parallel with most of the railway routes to be closed, the hardship occasioned will be slight—and also that the closure of branch lines will not greatly reduce

[1] D. L. Munby, Reader in Economics and Organization of Transport at Oxford University, quotes examples in the June 1963 issue of the *Journal of Industrial Economics*.

[2] Though an accurate comparison is impossible, the Beeching plan also seems to rely on a much higher rate of saving than the average achieved from closures so far. *The Reshaping of British Railways* states that a net saving of £18,000,000 will be achieved by withdrawing 68,000,000 passenger train miles, £11,000,000 to £13,000,000 by 'subsequent closures of lines and

policy of creating unstaffed halts, in which they contend that the only saving is in wages; that buildings, footbridges, and approach roads, still have to be maintained and lit; that only a limited number of tickets can be issued on trains even if the guards have the time to issue them at all, which they can do only on corridor stock; that children will play on the platforms, may fall under trains, and damage windows and lighting installations; that there is sometimes destruction caused by deliberate hooliganism if stations are left open and unattended.' This defeatism supports the contention that British Railways are uninterested in making the best of admittedly difficult circumstances. To state that 'the only saving is in wages' blurs the fact that at many stations wages account for 75 per cent of total costs; the reference to corridor stock is largely irrelevant as shortly all rural services will be formed of diesel multiple stock; and though there may be hooliganism in certain urban areas, unstaffed halts (with the minimum of buildings) have been maintained for years without damage or accident on many branches in the West Country and Scotland.

*Postscript*

traffic on the main lines.[1] This is perhaps the most pathetic of all the misconceptions about the rural transport position in recent years. Had the railways analysed their branch-line traffic more carefully, they would have realized long ago that where buses were convenient most public transport customers were already using them; and conversely that people still travelling by train belonged to a few well-defined categories for whom the ordinary bus would not suffice. It is said to be Dr. Beeching's own idea that the report should include a map of bus routes throughout the nation. Its production in a few days was a technical feat, but for the railway traveller it has no practical value. The long-distance passenger is not impressed by the assurance that in the fullness of time it would be possible to travel between almost any two villages in Great Britain: he wants to know what buses connect with trains starting from London and other big centres, and he wants to know their exact time before he starts his journey. As stated earlier in the book, in many districts the only hope probably lies in special railway buses—fast single-deckers with room for luggage, run as part of the railway system. British Railways have powers to introduce such services, yet the report does not deal with them.

### OTHER RAILWAY DEVELOPMENTS

The last six months have brought much talk but little action. The process of closing branch lines was positively halted during the final stages of the preparation of the Beeching plan. Further

reductions from passenger to freight standard of maintenance of other lines following the withdrawal of services and closure of stations', passenger services being withdrawn from 5,000 route miles, and further savings by closing many lines also to goods. The Central T.U.C.C. state in their *Annual Report* for 1962 that since 1950, 2,521¼ route miles were closed to passengers only, 796¼ to freight traffic only and 811 to passengers and freight. This reduction of about one-fifth of the passenger mileage of British Railways, plus the freight economies, yielded an estimated accumulated minimum annual saving of only £5,184,952.

[1] The £18,000,000 net saving from the withdrawal of stopping trains mentioned in the previous note is arrived at by taking the loss of revenue of £15,000,000 from the gross savings of £33,000,000. This £15,000,000 is made up of £12,000,000 in earnings on the services concerned, and a mere £3,000,000 in 'contributory revenue'.

*Postscript*

delay in effecting closures already 'in the pipeline' has been caused by the fact that under the new procedure (see page 42) the Minister of Transport has to deal with each case personally, and at least temporarily lacks adequate machinery to make the necessary investigations. Thus some little-used branches which should have closed years ago continue to operate at a heavy loss, and in several cases the extension of their life has necessitated capital expenditure on track and other renewals. There is still a desperate shortage of money for small capital works such as the elimination of unnecessary signal-boxes which might yield useful savings even if the branches concerned were to be closed within two years.

The new procedure governing the Transport Users' Consultative Committees has aroused grave misgivings, *The Guardian*, for example, devoting several leading articles to what it regarded as a serious loss of individual rights. The indications are that the T.U.C.C.s intend closely to follow their brief to consider hardship only. Questions of general economic prosperity, the size of traffic, and the cost of running a simplified service will not be discussed. Indeed, under the new procedure the committees will not even be furnished with a statement of the savings British Railways hope to achieve. The Central T.U.C.C. will thus be unable in future to publish in their *Annual Report* the savings so far accumulated. In calling attention to the last set of these figures they can present,[1] the Committee reiterate that 'the negative policy of closing down uneconomic facilities, while contributing a small financial saving, is not the panacea it has sometimes been made out to be. Each closure diverts some business to the roads. These losses of traffic must have contributed materially to the poor results of the last five years.'

While the T.U.C.C.s supposed to represent the public are unable to consider closures in their broader economic and social context, it must be added that a growing number of economists and others have become aware of the need for a more careful assessment of cost in relation to social benefit. The sections of the 1962 Transport Act dealing with branch-line closure procedures are in direct opposition to contemporary economic thought.

[1] The accumulated annual total saving of £5,184,952 mentioned in the earlier note in this chapter.

*Postscript*

### THE NORTH DEVON RAILWAY INQUIRY

In March 1963 the author was engaged by the North Devon Railway Action Committee to conduct an inquiry and to make recommendations. A report has been published,[1] but the chief findings may be briefly summarized here. Several of them vividly illustrate points made elsewhere in this book.

It was discovered that North Devon's railways are making an annual loss of £5 to £6 per head of the population served. Some trains are well patronized, but years ago many others could have been withdrawn, and some small stations closed, with only scant inconvenience ensuing. Fares cover a mere 2 per cent of the cost of running the passenger service on one branch line; on another line, one train carries more passengers than the other five of the day put together. In continuing to stop at numerous villages still without buses, the railways are not merely incurring prodigious losses, but running such slow and ineffective services that much of the more valuable medium- and long-distance traffic has disappeared.

Most stations should be closed, and a skeleton service of faster trains run primarily for long-distance passengers on the Exeter-Barnstaple-Ilfracombe and Barnstaple-Bideford sections. This, plus the subsidizing of minimum bus services to replace trains in certain villages, would involve a loss about one-sixth of the present size. A subsidy of £1,000 per mile of railway kept open for passengers 'would leave British Railways with a reasonable share of the burden, bearing in mind the feeder value of North Devon's traffic to the main lines'.

The position in North Devon bears witness to the weakness of the new T.U.C.C. procedure concerning objections to branch-line closures. It will be easy to prove that the withdrawal of little-used trains from hamlets without alternative public transport will cause hardship, although common sense demands that the railways cease handling this type of business. It will be extremely difficult, however, to show that the withdrawal of much busier long-distance services to Ilfracombe (which has ample *local* bus services) will seriously undermine the prosperity

[1] North Devon Railway Report (David & Charles, Dawlish, 1963).

*Postscript*

of a holiday area with a persistently high unemployment rate during the past decade.

North Devon bus connections shown in the railway timetable for many years have been removed from the summer 1963 issue—at the very time that sensible rail–road connections are being promised as part of the rationalization scheme.

BUS MATTERS

On the bus side, there have been few developments in the first half of 1963, the Government still shelving pleas for subsidies for rural buses. The only positive hint of any subsidy has been for new services needed to replace railways, but no opportunity has been lost to point out that buses already run parallel with most of the branch lines to be closed. As already stated, the false conclusion is drawn that not only will this mean little hardship when the trains go, but that the bus companies will be helped by extra patronage. In reality, in many areas most people who now travel by train will either use cars or not travel at all when the railways are closed.

Although the case for subsidizing buses would seem to have been amply proved by other investigations, including those of the Committee on Rural Bus Services which reported two years ago, early in 1963 the Ministry of Transport launched a series of inquiries into the transport position in different parts of the country. They quickly ran into trouble. In the case of the selected area in Mid-Devon, for example, it was soon found necessary to extend the boundary in order to collect worthwhile information, while many people refused to answer the questionnaires in the form chosen. Were such further inquiries necessary, the employment of a market research firm rather than the Ministry's own staff would have been preferable. Far more useful, however, would be an investigation into how different types of subsidy might work in practice in selected areas.

# INDEX

Agriculture, 5, 9, 68, 94, 95, 97–9; *see also* Forestry
Ashburton branch railway, 116, 138
Ashton, 89, 93–110, 111, 112, 126–35

Beeching, Dr. Richard, 4, 14, 166, 168
Bellingham, 93, 106
Birmingham & Midland Motor Omnibus Co. Ltd., 46
Birtley, 71, 89, 93–110
Blencow, 151, 152, 153, 154–5
Border Counties railway, 93, 116
Bovey Tracey, 115, 136, 140 n., 141, 142
Bridford, 129, 132 n., 133–4
Brigham, 150, 151, 154
British Electric Traction, 16, 17, 18, 45
Broughton, 144, 148–9
Bus companies, 3, 15–18, 45–51, 55–60, 78, 81, 133, 142–3, 157–65
Bus trailers, 54, 63

Canonteign, 112, 133
Carlisle, 150, 153, 156
Car ownership, 7, 18, 20, 67, 70–3, 74, 94–110, 112–19, 132, 135
Carriers, 11, 15
Chagford, 115, 135, 142
Christow, 106, 111, 112, 126–35

Chudleigh, 111, 112
Closures, railway, xii, 4, 5, 13–15, 21–6, 36–44, 73, 81, 111–16, 123, 128, 137, 140, 143–9, 150–7, 166–9
Cockermouth, 53, 151, 154, 155
Commuting, 30–2, 94, 99–105, 112–16, 127, 129, 132, 134–5, 137–8, 142, 146, 152–3
Competition, rail/road, 13, 16, 127, 135–7; *see also* Car ownership
Coniston branch railway, 116, 143–9
Connections, bus, 105–7, 111, 141–2, 160
rail, 35, 94, 116, 151, 153
rail/road, 32, 44, 52–5, 81, 84, 111, 113, 114, 132, 141, 146, 153, 168, 171
Contract buses, 56 n., 58–60, 61–2, 74, 76, 78, 79, 81, 106, 111, 132 n., 133, 145, 148, 158, 164–5
Coras Iompair Eireann, 22, 53
*Cornish Riviera Express*, 25
Councils, *see* Local authorities
Cross-subsidization, 4, 6, 17, 46, 47, 61, 76, 81, 83, 158, 164
Crosville Motors, 49 n.
Cumberland Motor Services Ltd., 46, 92

Dartington Hall Trustees, xi, xii
Depopulation, 8, 9, 18, 65–70, 73, 106

173

## Index

Devon General Omnibus Co. Ltd., 94, 106, 114, 127, 129, 132–5, 141–3
Dieselization, 22–6, 29–30, 113, 138–9, 140, 151
Doddiscombsleigh, 112, 133, 134 n.

Education authorities, 25, 58–60, 61–2, 82, 145, 165
Eire, 22, 53
Embleton, 151, 154, 155
Exeter, 94, 106, 111, 113, 126, 127, 128, 132–4
Exeter Railway Company, 126
Exe Valley branch railway, 22–3, 24, 32 n.

Fares, bus, 18–19, 20, 51, 52, 58–9, 116, 146
railway, 7, 167
Ferries, 73
Finance, bus, 19–20, 45–8, 54, 55–60, 146, 148, 157–65
railway, 14–15, 21–6, 32–3, 54, 128, 138–9, 144, 148, 149, 150, 154, 167
Forestry, 68, 73–4, 93, 94, 97–9, 110–11
Foxfield, 143–9
Fuel tax, 20, 55, 75, 76, 80, 82

Great Western Railway, 15, 16, 126, 135
*Guardian, The,* 169

Halts, railway, 22, 28, 34, 127, 155, 166
Heathfield, 111, 112, 115, 116, 126–35, 140 n., 142
Helston branch railway, 24–6
Hexham, 93, 105, 106, 111, 157, 160

Highland Transport Inquiry, 47, 56, 58, 59, 77, 78, 84
Highland Omnibus Co., 47
Hookway, R. J. S., 68 n.
Housing and Local Government, Ministry of, 74
Howe, H., 41 n., 43–4

Ilfracombe, 170

Jack Committee, *see* Rural Bus Services, Committee on
James, W. T., 76, 83

Keswick, 52–3, 116, 123, 150–7

Lake District Transport Inquiry, xi, xiii, 14, 46, 52–4, 70–1, 89–110, 116–25, 145–57
*Lakes Express,* 151, 156
Launceston, 23, 24
Level crossings, 33, 154, 155
Lewes—East Grinstead branch railway, 40
Licences, 17, 19, 48, 56 n., 61, 76, 77, 78, 81, 133
Licensing of Road Passenger Services, Committee on the, 17
Local authorities, 7, 50, 58–61, 73, 76, 77–8, 79, 144, 145, 147; *see also* Education authorities
London & North Western Railway, 13
London & South Western Railway, 126
London Traffic Act (1924), 17
Luggage, 54, 55, 63, 141, 145, 168
Lustleigh, 113, 114, 115, 116, 136, 141, 142

MacBrayne bus services, 47
Mails conveyance, *see* Post Office
Mennear, Douglas, 157
Merioneth, 59
Mills, G., 41 n., 43-4
Mini-buses, 62, 135, 165
Moretonhampstead branch railway, 113-16, 135-43
bus, 129-33
Motor cycles, 94-105, 135
Munby, D. L., 167

Nationalized Industries, Select Committee on, 6, 13
Newton Abbot, 94, 106, 111, 113, 114, 115, 126-9, 132, 135-43
Nicholas, H. R., 76, 78
North Devon Railway Inquiry, 170
North Eastern Railway, 27
*Northumberland Country Bus*, 19 n., 58, 157 n.
Northumberland Rural Community Council, 19 n., 157

Okehampton, 142
One-man buses, 62

Penrith, 46, 53, 116, 150-7
Plymouth, Tavistock & Launceston branch railway, 23-4
Post Office, 25, 62-3, 82, 148
buses, 62-3
Powley, E. B., 76-7, 78
Prams, 107, 116
Public relations, bus companies, 45, 49-51, 61, 73
railways, 36-7, 73, 113, 125, 126, 127, 137, 157 n.

Public Transport Association, 47, 48

Quarrying, 93, 94, 97-8, 101, 127

Railway Board, 18, 39
Railway buses, 15-16, 23, 43, 53-5, 84, 111, 113-14, 128, 129, 132-5, 142, 145-6, 168

Ravenglass, 155
*Reshaping of British Railways, The*, 166, 167 n., 168
Ribble Motor Services Ltd., 46, 49, 145-7
Riccarton Junction, 93
Road Traffic Act (1930), 16, 17
Rural Bus Services, Committee on, xii, 5, 19, 20, 46, 47, 51, 56 n., 59, 69, 75-9, 82 n., 83, 171
*Rural Transport: A Report*, xii, 15, 75, 78, 80
*Rural Transport Problem in Wales, Report on the, see* Wales and Monmouthshire, Council for

Saville, John, xii
School buses, *see* Contract buses
Scottish Omnibus Group, 17, 47
Services, bus, 46-7, 53-4, 93, 94, 106, 110, 111-13, 129, 132-5; *see also* Railway buses
train, 30-2, 52-3, 94, 127-9, 135-8, 143, 151-2, 154-7
Shelters, bus, 55, 63, 106, 113, 116
Shopping, 9, 18, 67, 69, 99, 105, 106, 113, 114, 120

## Index

Shops, mobile, 18, 106, 114
Signalling, 22, 31, 32–3, 84, 128–9, 136, 139, 154–5
South Devon Railway, 135
Staff, bus, 3, 48–9, 51, 57, 62, 164
  railway, 3, 14, 27–8, 33–4, 128–9, 139, 154–5, 157
Stage-service buses, 19, 58–9, 62, 63, 75, 158
Stations, railway, 12, 34, 125, 166; *see also* Halts, railway
Stonehaugh, 74, 110–11
Subsidy, 4–10, 14, 15, 44, 55–60, 65–70, 75–85, 170, 171; *see also* Railway buses

Tavistock, 23–4
Taxis, 59, 69, 105, 106, 114, 134, 135, 141
Teign Valley branch railway, 111–13, 116, 126–35, 140 n., 142
Teign Village, 112, 133
Tendering, *see* Contract buses
Thesiger Committee, 17
Threlkeld, 150, 154
Tilling Group, 16, 17, 18, 45–6, 47, 50, 82 n.
Timetables, bus, 63–4, 160–1, 164; *see also* Services, bus
  railway, 37, 55, 63–4, 132, 136, 137, 141, 146, 171; *see also* Services, train
Torbay, 113, 126, 135, 137, 138, 139, 140
Torver, 144, 147
Tourism, 18, 25, 68, 101, 116, 117, 120–5, 135, 137, 141, 143, 151–3, 155–6, 160–1

Town and Country Planning, 9–10, 73–4, 110
Town Planning Institute, *Journal of the*, 68 n.
Traffic Commissioners, 17, 48, 50, 55 n., 57, 58, 59, 60–1, 62, 63, 75, 76–7, 78–82, 133, 163–4
Transport Act (1962), 18, 39, 45, 169
Transport Commission, 3, 13, 18, 38–9, 43, 46
Transport Holding Company, 18, 45–6
Transport, Ministry of, 14, 27, 38–9, 75, 80, 81, 82, 140, 157, 169, 171
Transport Users' Consultative Committees, 13, 36–44, 48, 75, 80, 82, 140, 143–5, 166–9, 170
Travel habits, 18–19, 30, 67, and Part Two
Trusham, 89, 93–110, 111, 112, 127–35

Ulverston, 145–7
United Automobile Services, 164

Wales and Monmouthshire, Council for, 47, 51, 54–5, 56, 59, 65–7, 72, 77
Wark, 74, 93, 110, 111
Westerham branch railway, 39
Westmorland Federation of Women's Institutes, 108–10
Windermere, 123, 145
Women's Institutes, 7, 108–10
Workington, 150–7

For Product Safety Concerns and Information please contact our EU
representative GPSR@taylorandfrancis.com
Taylor & Francis Verlag GmbH, Kaufingerstraße 24, 80331 München, Germany

www.ingramcontent.com/pod-product-compliance
Lightning Source LLC
Chambersburg PA
CBHW070613300426
44113CB00010B/1513